MW01641236

Evolution

Fact, Fiction, or Fancy

Evolution

Fact, Fiction, or Fancy

Bernard J. Ficarra, M.D.

Rutledge Books, Inc.

Danbury, CT

Rutledge Books, Inc.
107 Mill Plain Road, Danbury, CT 06811
1-800-278-8533
www.rutledgebooks.com

Manufactured in the United States of America

Cataloging in Publication Data
Ficarra, Bernard J.

Evolution: Fact, Fiction, or Fancy

ISBN: 1-58244-166-9

1. Evolution -- Charles Darwin -- The Origin of the Species. 2. Historical Treatise. 3. Propaganda and Public Relations. 4. Scientific Methodology.

Library of Congress Catalog Card Number: 2001094469

— CONTENTS —

— CONTENTS —

— CONTENTS —

To all the ultrafine
American,
Austrian, British,
Canadian, French, Italian,
and
South American scientists
I have known during my lifetime.

Chapter 1

Scientific Setting in the Nineteenth Century

Most people say it is the intellect which makes a great scientist. They are wrong: it is the character.
—Albert Einstein (1879–1955)

Conceivably the twenty-first century will usher into reality a decrease in the growth and enlargement of scientific enterprises. Scientific knowledge will not cease to expand. But the growth of the profession of science as a scientific, federally funded undertaking will diminish (Goodstein 1995). Biotechnology will not fall victim to any downsizing. Federal policymakers and bioscientists consider biomedical research a necessarily important means of strengthening the United States of America's high-technology competitiveness (Kevles 1995).

In the nineteenth century, sciences in general had a burst of growth, value, astonishment, and willing audiences. Even Johann Wolfgang Von Goethe (1749–1832), the creative poet and Germanic dramatist, was a pioneer in projecting the concept of evolution. His knowledge of biology was acutely profound. It is not proven historically, but he is alleged to have been the first to coin and use the word *morphology* (Himmelfarb 1959). Thus the stage was prepared for the arrival of evolution as a revolutionary concept in bioscience. A ripple effect commenced whose waves have continued to the final hours of the twentieth century and the commencement of the twenty-first century. Without much query,

Darwinism flourished in certain, specific scientific circles. This provocative theory goes beyond standard scientific norms as it considers the human context in which it originated. A wider understanding was attempted than was conceived before Darwinian evolution was postulated. This unique perspective did not augment science's place in human culture. Many formal scientists, willingly or unwillingly, were ensnared into the discourses on either the overvalue or the undervalue of Darwinism as a bioscientific certitude.

Responsible behavior of scientists in either advocacy or negation of a proposed dictum must be with moderation and adherence to proven, discernible truth. Beyond scientific recognition in the marvelous cosmos, thoughtful scientists must appreciate their singular positions as intellectual leaders and adherents to moral truth and ethical performance (Polkinghorne 1996).

In the science wars of those who propose theories and their opposing critics, objectivity may be missing. Sociologists, philosophers, theologians, and many others in high-rise disciplines degrade scientists for unstable assumptions. The criticism is dismissed cursorily as bioscientists ignore the historical requisite of objectivity in science (Bower 1998).

Unless this is done misunderstandings in science will increase to its own detriment. Then a nonscientific reconstruction of science and technology might occur. Necessity for correction will delay advancements in diverse fields of science. This will require the offering of a new perspective for the public's understanding of science. Clarification will include challenging existing ideas on science in relationship to humanity and sociology. Highly relevant concern may focus on evolution over its resurgence in popularity, its scientific authority for credibility, and its effectiveness as a theory worthy of serious consideration. The nonscientific sector of human society must not be doubtful of its commitment to truth, which should be an essential constituent of all science (Irwin and Wynne 1996).

Since the eighteenth century when Samuel Johnson,* called *Doctor* because of his erudition and lexicography, the noun *evolution* meant a gradual process in which something changes into a significantly different thing. In the nineteenth century the appearance of Charles Darwin (1809–1882) on the illuminated stage of science gave a new definition to the word that seems to have diminished any other use thereof. The change occurred when the British naturalist expounded his theory of evolution by natural selection in his 1859 publication, the *Origin of Species.*

Bioscientific usage gives preference to the definition that stresses the theory of progressive metamorphosis. More succinctly it is a proposed theory on the philology of organisms. It predicates a belief that groups of organisms, as species, may change with passage of time so that descendants differ morphologically and physiologically from their ancestors. Intellection gives credence to those who state the theory is pregnable, not sustained by facts, and is lacking in substance. Evolution as a proposed science is not wide, accurate, and solid. It is not a foundation upon which additional doctrines are provable. In itself, evolution as a theory is not spectacularly overwhelming to the degree that it eclipses other bioscientific accomplishments of the nineteenth century.

The pathway of keen research is not synonymous with an intellectual process that engenders a theory, a hypothesis, and/or a speculation. Educational studies that enhance scientific pursuits do not stifle originality of thought. Contrarily, such formal learning is the essence of initiative thinking that instigates creativity. Often in the guise of mental curiosity the fascination for scientific research does not always lead to worthy discoveries. However, intellectual inquisitiveness is not to be ignored. It permeates, even haunts, the scientific world and allied professions.

This attribute has been noticed in many bioscientists. Some

---

* Asterisks througout text indicate explanation found in glossary.

have come forth as Darwinian advocates who defend and foster evolution in various interpretative ways. The evolutionary process has been applied to disciplines beyond biology. Nowadays it is nearing a philosophical concept which some adherents apply to subjects not associated with bioscience.

The specific word *evolution* has been applied to evolutionary psychology and other disciplines. A psychologist has written that the critical evolution of the past million years has been the evolution of the mind. This is maintained with the statement that "psychology has long been deeply ambivalent about Darwin's unsettling discoveries." The author describes a new rapprochement which is termed *evolutionary psychology.* It is surprising to him, even astonishing, that Darwinism could have been disregarded by academic psychologists. Adamantly critical of colleagues, a demand is made that the relationship between the two (psychology and evolution) must be readdressed. It is postulated that "an evolutionary perspective helps us understand what we are and how we got that way" (Plotkin 1998).

A philosopher and a biologist united to demonstrate to their satisfaction that unselfish behavior is in fact an important feature of both biological and human nature. Their combined effort was to prove once and for all to those interested in this subject that behavior can evolve by natural selection. To their many colleagues credence has been given for their assimilation that group selection has minimal controversy. By means of a detailed case study they advance what is termed "an indisputable argument for group selection as a legitimate theory in evolutionary biology" (Sober and Wilson 1998).

Chapter 2

Dinosaurs and Birds

The postulants of evolution in its applied modifications are not to be minimized, degraded, nor ignored. They have a massive amount of information with procedural reasoning that reinforces support that is overwhelming (Rodenhauser 1996). On the doubter's side there are anti-evolutionists who do not tolerate uncertainty well. They question new theories readily and more easily are suspicious of unproven existing theories (Knight 1982).

Enthusiasm for evolution is not lukewarm in this century. Modern researchers introduced a 260-million-year-old fossil that they say bolsters the theory that South Africa was the place where scaly, cold-blooded reptiles evolved into the ancestors of furry, warm-blooded mammals. This shows that South Africa is the home of a specific dinosaur (Modesto 1999). The fossil is claimed to be a reptile-mammal link with features of both species. Found in 1995 near Williston in the Karoo, it was named "anomodont, the lawless-headed one of Africa." Finds of such dinosaurs in South Africa have led Sean Modesto (from Toronto, Canada) and paleontologist Bruce Rubidge, of the University of Witwatersrand in Johannesburg, to conclude, "They were once plentiful there and provided a good food supply for local predators" (*Proceedings of the Royal Society of London,* February 1999).

Until recent years, scientists had believed this species originated in Russia. At that time, geologically, there was one supercontinent, known as Pangaea. James Hopson, an expert on anomodonts from the University of Chicago, said the find was not earthshaking but reinforced the growing acceptance that anomodonts originated in South Africa. He said the fossil was in poor shape and he questioned its age (Johannesburg AP dispatch 1999). The reptile-mammal relationship received no formal comment from the Darwinian cohort as it continues not to have scientific documentation.

With concision it may be stated that sideline-educated spectators accept evolution as the source for unending debate, not as dogma. Truth and falsity tug at the conscience of scientists as contemplation over new proposals arouses skepticism. Non-strident, negative denials of the reptile-mammal connection without intention categorize it as a fanciful assertion, which adds to the belief that evolution is a losing theory.

A sensational and radical theory has turned the world view of most paleontologists upside down. It is the hypothesis that "dinosaurs gave rise to birds" (Martin 1998). The dinosaurian origin of birds is not original with Martin. "It was first put forward in 1868 by the English biologist, Thomas Henry Huxley" (Martin 1998). He advocated this belief with fluted hubris.

Huxley's idea dissolved out of favor when it was replaced by the proposal "that birds evolved from reptiles that lived before the dinosaurs" (Ostrom 1998). No nondinosaurian reptile has been identified as even a possible bird ancestor . . . the basis for the bird-dinosaur relationship has been anatomic evidence regarding hands. Martin writes, "as if it had already been proved that the fingers in certain dinosaurs and birds were homologous." But there is a lack of confirmation. The paucity of directly comparable evidence that could prove "bird hands and dinosaur hands are topologically distinct" has not eliminated the bird-dinosaur theory (Ostrom 1998). Such promulgations are made by those who see and hear only what contributes to personal ideas of the moment. No matter how well they are stated they remain flawed.

Hilarious as it may seem, the hypothesis that birds evolved

from small carnivorous dinosaurs is not a caricature to some bioscientists. They argue that the opponents of the dinosaurian origin of birds do not agree on a common objection. Moreover, they say that no specific alternative has been proposed to negate the concept. Antagonists have disagreed without offering a testable alternative (Padian 1998). Insightful review may elicit vigorous denial of this thesis. Contrarily, some paleontologists who maintain the vista that birds were derived from advanced theropod dinosaurs (maniraptorans) have much support. Denouncing and supporting literature on this controversial subject is not lacking as it adds simultaneously to the complexity (Wilson, Ruben, and Padian 1998).

More is being spoken about the current pursuit of phylogenetic trees, geologic clocks, and exceptions in the Darwinian theory. Few species of animals or plants reproduce asexually and those that do seldom comprise an entire genus. There is scant evidence that sexual reproduction plays an essential role in evolution (Meselson 1998).

A compilation of thinkers responsible for the modern synthesis of evolutionary biology and genetics published their updated thoughts in one volume. With a new preface and alert editorial supervision, an analysis of evolution is remade. This publication "calls attention to the fact that scientists in different biological disciplines varied considerably in their degree of acceptance of Darwin's theories" (Mayr 1998; Provine 1998). Confusion continues to reign in the disunity among bioscientists, even in those enamored of evolution with effusive sentimentality. Some scribes on evolution are not always irreproachable as to adverse vociferous discord among themselves on the theory they supposedly relish with fond regard.

Without abandonment of skills while immersed in queries and inquiries, retention of imaginative brilliance is not disturbed in researchers of merit. Applied to scientists, their activity under either similar or like circumstances propels them to efforts that advance all categories of research. Different answers to many questions eventuate out of the dedication expended in carefully thought-out research projects. Opinions arise that either affirm or deny proposed inventions, discoveries, and theories. The conclusions reached may "accept, transcend, or modify reality, perhaps to introduce a new version of reality, to make the ordinary extraordinary, and to advance knowledge, culture, and humanism" (Rodenhauser 1996).

Chapter 3

Specimen Collection

Joseph Banks and Charles Darwin were members of Britain's elite. Born into a world of dynamic marriages, they possessed an automatic entrée into academic circles. Other, less advantaged men gambled that by going to the ends of the earth they would acquire valuable social connections—and in fact the rewards could be great. Scientific travelers received gold medals, had their portraits painted, and wrote popular memoirs. Thomas Henry Huxley, born over a butcher's shop in Ealing, England, found after his journey to Australia and New Guinea on the HMS *Rattlesnake* that the doors to the scientific community were not closed to him (Rittimann 1998).

Added to the prestige was a sizable income derived from specimens gathered during voyages. These were curiosities in much demand which sold at high prices to collectors and professional exhibitors who owned the displays in their curiosity cabinets. Collections appeared in major European cities. Privately owned, each display was a lucrative business. Money was derived from visitors who paid to enter the curiosity cabinets.

In such collections all kinds of inanimate and animate objects were exhibited. Some were fragments of animals that left much to the viewer's imagination. The proprietors of these collections "were princes, naturalists, apothecaries, and ambitious businessmen" (Rittimann 1998). All became wealthy in various degrees. Some exhibited living people with deformities and/or other anatomical anomalies. This was the inspiration for future circus formations with sideshows.

Such curiosity cabinets were the forerunners of today's well endowed natural history museums. They were born in an age of exploration when evidence of exotic people, plants, and animals was beginning to challenge the wisdom of humanity. It deepened when, in 1859, Darwin published the *Origin of Species*. Until then many naturalists and most nonscientists accepted the biblical record that God had created animals, each having a perfect and unchanging form. Suddenly, the animal kingdom is to be a much more haphazard affair. Under Darwinism, species developed in response to their environments and change was the norm. Darwin's shocking new theory implied that human beings were not categorically different from other animals. People descended from apes? "It was as if someone had placed fun-house mirrors inside the Garden of Eden" (Rittimann 1998).

Darwinian concepts stimulated an enormous fascination with the bizarre. Victorian England was captivated by abnormalities. High markets developed in human ugliness. People came to gaze at giants, dwarfs, partially conjoined infants, and Siamese twins. "People who seemed to possess monkey-like characteristics were particularly in demand" (Rittimann 1997).

A Laotian girl was exhibited in 1883 as Darwin's missing link because she was unusually hairy and could pout like a chimpanzee. The foremost attraction was her hirsutism. Spectacles such as the "Hottentot Venus," an African woman named Sartijie Baartman with unusually large buttocks reminiscent of the genital swellings of female baboons, and the bearded Mexican woman, Julia Pastrana (advertised by one impresario as part human, part orangutan), also challenged the boundaries between human and ape (Ritvo 1997; Rittimann 1997).

Such obsession with the unconventional, the grotesque, the monstrous, the exotic, and almost macabre exhibits did not originate during the Victorian age. Written documentation indicates that such fascination dates back to the beginning of recorded history. The ancient Greek historian Herodotus wrote of ants the size of dogs. Scholars in the Middle Ages told of Chinese trees whose fruit contained miniature sheep. The birth of conjoined twins near Florence in 1317 startled the Europeans as it was viewed as an evil omen for the entire city. During the halcyon days of European exploration, returning sailors told tales of monstrous tribes and many bizarre, grotesque things (Purcell 1997).

Many European naturalists tried to make sense of the new arrivals; several of their more inquisitive colleagues ventured into any known strange wilderness. In the eighteenth and nineteenth centuries Joseph Banks, Charles Darwin, and Alfred Russell Wallace, among others, set sail for Africa, the Amazon, Australia, and Southeast Asia, with a single, monumental aim: to discover and to name every living thing. The naturalists trapped, stuffed, and pickled their way through the tropics. Some added cooking to that list. They had eaten themselves into the animal kingdom further than any other human being (Raby 1997).

"The duck-billed platypus stumped nineteenth-century naturalists. When that odd-looking creature was first brought to England from Australia, no one could figure out what category to put it in" (Rittiman 1998).

Now the platypus is classified as a monotreme. It is one of three primitive mammalian species that lay eggs. To early zoological observers it seemed part bird, part fish, and part quadruped. It was like a fanciful illustration from a medieval bestiary. Some original investigators suspected it was a fraudulent mixture, a human-made concoction. Therefore they checked closely to see whether the platypus's flattened beak had been sewn on or was real (Ritvo 1997). In the realm of the *Origin of Species*, the platypus has not been catalogued precisely.

The fascinating lure of otherness is undeniably present to this day as it juxtaposes fact and fiction. Often the differentiation between the real and fake is not clear-cut. Culture, science, fiction, and myth at times are mixed in a potpourri that is unfathomable. Truth is confined to a limited realm. But lies are extensive and not restricted. Darwinism did not elucidate the difference but added to the confusion.

Chapter 4

Evolution is Popular

Discussions on evolution are becoming more frequent. Some commence as a philosophic presentation and can terminate in a bioscientific discourse. Opposite to that, an intellectual session could start with evolution as a biomedical topic and find the disputants speaking like philosophers. Darwinism is becoming the fountainhead of remarks in social, scientific, and spiritual circles. Mental concentration attempts to ascertain all that has been, is, and can be learned about nature and religion under the light of evolution.

Some scientists are creating a new discipline in the biosciences with many assumptions based upon Darwinism. Their efforts are not to be minimized. A new discipline may be the first rock in the building of a milestone for a scientific pathway leading to greater understanding of biological phenomena. Even with hoped-for benefits, the more scholars learn, the greater becomes the realization that there are intellectual limitations to final answers. Only the unseen future will witness the probability that noumenal boundaries will be pierced.

The doctrine of evolution variously applied in different ways by individual scholars results in many interpretations. Hence it is deduced that it is inadequate to satisfy the questioning intelligence of many academically oriented bioscientists, philosophers, and other persons of similar professional status. On the subject of evolution promulgated in the nineteenth century, it appears to have had a reawakening in the twentieth century and continued into the twenty-first century. With this renewal of interest, the topic has acquired a persuasive influence which prompts university faculties to teach it anew. Men/women of marked intellectual inquisitiveness have disenthralled themselves from prejudicial thinking. With open-mindedness, minus overpowering outside influence, Christian scientists, in their own fashions, have presented their views and opinions on the generic theme of evolution.

Attempts at solving the aging puzzle has attracted zoologists and gerontologists to seek answers to the decaying process of growing old. Researchers are turning to evolution, which may give a clue as to why living things, including human beings, age at diverse rates. This quest has engaged biologists for many decades. They have advanced some three hundred and thirty different theories of aging to explain the happenings in body deterioration (Stewart 1998).

No evolutionary pressure to accelerate the wear-out pathology is identifiable. "Fast-aging, quick-dying animals like opossums and mice have no special protection. They have evolved the ability to breed early and plentifully, before an accident kills them" (Austad 1998). This is in keeping with the evolutionary theory of aging proposed by the English biologist Peter Medawar and others (Stewart 1998).

During the 1950s the Medawar theory was overlooked, minimized, and ignored by disparate gerontologists and bioscientists. "The basic idea is that the force of natural selection has great power to favor good genes and act against bad genes" (Medawar 1950).

Austad says, "If we all had genes that killed us on our one hundredth birthday, it would not matter very much. On the other hand, if there was a death gene that switched on at the age eleven, almost no one who carried it would live long enough to reproduce." The gene would rarely be inherited; it would likely appear only as a mutation. "Austad made news headlines in the gerontology world. A hurdle for the evolutionary theory of aging, he knew, was the difficulty of comparing aging rates across species—there were too many differences from one species to another" (Stewart 1998).

Austad's view of aging as a byproduct of Darwinian strategy, not as a biological mishap, is a perspective he and others had been trying to persuade scientists to believe and accept (Stewart 1998). The influence of Darwinism is not stagnant. Intellectually mobile, it marches on into the ranks of eager bioresearchers. The current evolution advocates are concentrating on geriatric questions. Seeking answers, investigations are being made in the nonhuman population. Disclosed studies reveal that an inherited trait allows some mammals to age rapidly and others to grow old slowly, according to the evolutionary process. The Gordian knot to unravel is the linkage, if any, to the *homo sapiens* of those theoretical observations allegedly attributed to animals.

Evolutionary concepts teach that no matter how much a theory may sound plausible, that does not *ipso facto* make it a dogmatic reality. Contradicting criticism by bioscientists, creation scientists, and philosophers does not deter non-pavid scientists from presenting propositions as factual that in their own minds they believe to be untrue. What is insidious about such proposals is that someone tries to incorporate it into the lives of others. Over periods of time what is proposed becomes accepted with dependency upon it for self-comfort.

Darwin's influence has been continual even to the beginning years of the twenty-first century. Currently, some university faculty scholars have developed a new biological subdivision on the evolution of aging. They posit, "When an animal has relatively few predators natural selection appears to exert a force that can cause the animal to age very slowly and evolve a long life span, particularly if it's a non-mammal" (Finch 1998).

Convincing documentation exists for a number of vertebrate

species that live for a long time, far beyond the age of menopause, without showing any decline in fertility. These animals are the basis for a new thinking about aging (Finch 1998).

In the 1970s, Caleb Finch made a radical proposal about the evolution of aging. Regarding the earliest vertebrates (animals with a spinal column: the back bones), from which mammals eventually arose, Finch suggested they had negligible senescence—that is, they either did not age at all or they aged so slowly that the process was almost insignificant. Finch said, as a universal feature of mammals, it is a relatively recent evolutionary invention. If Finch's idea is right, many non-mammalian species living today should have retained the ancient characteristic of having a negligible senescence (Montgomery 1998).

The best-studied example of an animal with negligible senescence is the Blanding's turtle. By analyzing life tables of mark-and-recapture turtles, a determination was made of the mortality rates for repeatedly released, marked turtles. Investigators found that turtles in their forties, fifties, and late sixties had the same mortality risk as the younger turtles. According to researchers, age-related changes are only biologically significant if they either impair fertility or increase the risk of dying. The Blanding's turtles, in this sense, show no signs of aging. They appear to live their entire adult lives on a mortality plateau where the risk of dying does not increase with the passage of time. Death of turtles is due to random accidents and bacterial/viral infections. Unlike human beings, the turtle immune system does not grow less efficient with advancing age (Congdon 1998).

In evolutionary terms, the purpose of life is to propagate genes by reproduction. Natural selection has less and less reason to keep persons healthy and alive as they become infertile and therefore less capable of reproducing. By age eighty, all women and most men cease to be fertile. At that age, the chance of dying is 50 percent, or one in two (Montgomery 1998).

Mortality figures are the highest between the ages of fifteen and eighty. The geriatric mortality rate shows a spectacular acceleration. But at the ages between eighty and eighty-five, a remarkable thing seems to happen. The mortality rate is sort of consistent. A mortality plateau is reached—and on a plateau, a person can still die from any number of causes, but the person will no longer be progressively deteriorating (Rose 1998). Researchers believe that such mortality plateaus are a universal feature of all organisms. According to this theory, when natural selection is through with a person, the plateau will go on forever until natural death (Rose and Mueller 1998).

Exploration of evolution along the pathway to etiology has taken some investigators down a much neglected avenue that links natural selection and genetics—the effect of changes to the rates and timing of growth and development. One researcher delves into the living and fossil worlds to show how animals and plants have evolved when the carefully orchestrated pattern of embryological development is gently nudged off-course. Demonstrated is how this phenomenon—known as heterochrony—has affected many aspects of evolution, including the mechanism behind the selection of different breeds of animals, differences between sexes, and animal behavior. Heterochrony explains how the dinosaurs got so big, how pterosaurs managed to produce a wing supported only by their fourth fingers, and what has driven the primate species with the biggest brain and longest childhood (McNamara 1997).

Genetic causation alone is not a primordial immunity against error and misjudgment. Comprehensive challenges recur to the unsustainable evolution development as posited. Even with much speculation, no integrative affiliation of physiological actions has demonstrated an emerging pattern of a sustainable planned reticulation for evolution. So many proposals have been made by wrongheaded scientists that high waves have resulted on the solemn sea of pensive thoughts. Confusion has increased. Like shipwrecked passengers in a lifeboat tossed by the tempest sea, they vacillate between being seaworthy and seasick. Tabling evolution as a disturbing subject is sometimes the only means of calming troubled waters.

Chapter 5

Charles Robert Darwin

Considered by many as a most distinguished scientist, Charles Darwin achieved much publicity. No doubt he used his deep insight with partial originality to increase productive thought among his domestic and foreign colleagues. His attractive and unusual biography has been written many times without unanimity of his status as a productive scientist of outstanding merit.

Added to his other bioscientific endeavors, Darwin examined the causes, both somatic and psychic, of all the fundamental emotions in man and animals. His conclusion was that, "The chief expressive actions exhibited by man and by the lower animals are innate or inherited." He believed that most of the movements of expression must have been acquired gradually over a period of time during an evolutionary process (Darwin 1873, 1898).

A study of his voluminous correspondence unfolds much of his personality and accomplishments. The documents describe his involvement with wider scientific endeavors. The first publication that brought scientific recognition to his name was in 1862. These were his two botanical papers and a book on the pollination mechanisms of orchids. More particularly, notice was given to the extent and breadth of the botanical experiments he had performed. In his letters, he writes of his scientific progress, but dispersed among them are references to his health. These recordings indicate the sad effects of Darwin's continuing ill health and the sorrow he endured from the serious illnesses of two of his children (Burkhardt et al.

1997). Medical problems in his household made him a periodic invalid (Colp 1977).

Some reviewers of Darwinian letters extol him. A selection of these writings from 1825 to 1859 are considered among the best. These letters reveal an intimacy of the person as he gives birth to significant ideas in the history of science. They are a private window through which the reader sees these epochal historical events that developed into the theory of evolution (Burkhardt 1996). These intimate self-expressions in writing, standing as literature, may be considered as a major work large in scope and splendid in execution.

Interpretation of Darwin's life is a study that combines biography and cultural history. An emphasis on the social impact of Darwin's work indicates that his contemporaries did not totally appreciate his thinking that subsequently caused turmoil among scientists and theologians. With his paternal inheritance on the evolutionary process he made himself a product of his own time. But he transcended his era by an idea that is being exploited to this day by many of the intelligentsia who disagree with his tenets (Bowler 1996).

At times entertaining, not always convincing, sometimes authoritative without despotism, Darwin's writings as an entity are a revelation of his humanity. His flair for English prose is enviable. Imagination is not restrained. Hard-headedness is mild. Clarity of expression is among the best. Combativeness is not dominating. All reveal a penetrating intelligence worthy of a career author.

Overall, however, his writings do not sustain proof, arguments, and scientific continuity that are sound. His theory of evolution does not comport with religious belief like other bioscientific systems. However meritorious his presentation on the origin of species may be as a sensational attraction, it has a spurious tinge. Additionally, beneath the unfailing calm tone of the theory, reasoning moralists discern a tone of religious disbelief. The question may be asked if Darwin was an agnostic, an atheist, or simply a nontheist.

There was nothing providential in the ideas of Darwin. On his own initiative he reworded, emendated, and added his personal touch to a previous concept expounded by his father and grandfather. His efforts perhaps would label him as an advocate of neo-scientific liberalism, not an originator of neogenesis.* This neologism is not sybaritic.** Darwin's ambition was not disclosed, but it may be inferred that he enjoyed the luxurious pleasure associated with fame. Inwardly he was anxious to have the sublime splendor derived from a new discovery. Objectively, as compared to brilliant scientists, he was not an indefatigable achiever.

His thoughts on evolution and the desire to prove it brought together in one person the urgency to accomplish the impossible. The dream was so intense he was driven to propose an innovative biological concept. With malleable integrity plus synergy of thought, his anticipated success was considered favorable by a personal assurance instigated by himself.

In an extraordinary, journalistic, brief writing entitled the *Most Evil Person of the Millennium,* Darwin was mentioned. Others considered were Stalin and Hitler. At Oxford University, Fernandez-Armesto chose Charles Darwin. Although considered to be a good and great man, he is accused as the propagator of evil. Perhaps unintentionally he "provided the appearance of scientific justification for the pernicious notions that violence is a social good (fight for survival) and that one man can be inherently superior to another simply by virtue of race" (Fisher 1995). Thus Darwin's postulate is categorized as a pseudo-scientific theory like that presented by the apogee of evil, Houston Stewart Chamberlain (1855–1927).

As the son of a British admiral, Chamberlain had exposure to the public. Thus he found a listening audience without difficulty. This scion of English society firmly believed in Nordic supremacy. Hence he was a strong influence on Nazi Germany. Flaming this fire of hatred, he increased the momentum that hastened World War I. After the Allied victory, this falsity did not vanish—the credo continued. It found support as it fomented hatred in Germany, America, and Arabia, culminating in World War II (Fisher 1995).

Such self-styled do-gooders are in reality evildoers. Relying upon their own inflated intellect, they are misguided beyond their own individual imagination. By such serious misalignment intellectualism is misdefined. Proper Oxonian usage categorizes intellectualism as the "doctrine that knowledge is either wholly or mainly derived from pure reason" *(Oxford English Dictionary).* It would be unkind as well as a rash judgment to indict Darwin as lacking in pure reason, but it is not improper to analyze his writings for logical deductions. He does not utilize syllogistic reasoning in arriving at his speculations, which are founded upon surmised notions. These in their own circumstances may be harmless; however, when a hypothesis importunes with impunity the direct creation of mankind then mental belligerency becomes rampant. Even with his criticism no maligning epithet has been applied either to him or his name. A terse example of a Darwinian epigram is: "A man who is writing about that which he knows nothing is absolutely invulnerable."

Darwin's finishing touches to his research thoughts on the origin of species were inspired when he visited and explored the Galapagos islands. To this day, from a distance they look bleak, barren, haunting, and mysterious. Just as in 1835, this archipelago off the western coast of Ecuador is primordial without many people but abundant with wildlife.

The water, white sandy beaches, and black massive rocks are a haven for crabs who scuttle freely over the sand. They go into the blue water and return to hurry across the sand to find a cozy resting place among the dark, clean rocks with sleek crevices. Here, too, are found iguanas, the only seagoing lizards in the world. They seem to know their indigenous recognition as rulers of their own domain.

The islands are among the best breeding grounds for sea lions.

Shallow water, yet sufficiently deep, protects the baby sea lions from sharks and whales. Shoals are never empty of infants who thrive and grow in nature's cradle. Close by, mockingbirds sing them to rest. These birds are voluminous. Rarely are they seen singularly but gather in groups of eight or more. Eerily, the islands are populated by only two human beings.

What Darwin wrote in 1835 is applicable in 2001. Nothing could be less inviting than the first appearance of the rough, barren landscape of the islands. It is a pristine preserve, a true animal kingdom. This feeling persists. One never ceases to have the sensation that this geography is either the end of the world or the beginning of it.

Out of the smoldering ashes of antiquity come the reminiscences of dinosaurs, unknown species, predators, and imagined similar creatures. All these scenes and things unseen are conducive to thoughts of natural selection, evolution, and the origin of species. An ambiance such as this gave firm credence to Darwin's convictions that had been instilled in him by his father and grandfather.

The more islands he visited, the stronger became Darwin's belief in his theory. Although he did not visit all the one hundred-thirty-odd volcanic and rocky islands six hundred fifty miles off the coast of Ecuador, those he did visit were wildlife havens. It was a thrill to see animals in their raw habitats. Darwin did visit the largest and best known of the islands, Fernandina and Isabella, as well as Española, a smaller but beautiful island known as Hood. Ninety-seven percent of the Galapagos is Ecuadorian National Park. San Cristobàl and Santa Cruz islands are populated with 16,000 natives.

Down House is the name of the Darwin Museum in Kent, which was the home of Charles Darwin. This was where he wrote the controversial treatise, *On the Origin of Species by Means of Natural Selection.* The house is located off Route A-233. Open all year except for the Christmas Season and New Year's Day, it has many visitors. When closed, two wrought-iron gates, cemented into strong brick pillars with large stone balls on top, obstruct admission to the property.

From the outside, the appearance of the building is that of a rambling mansion. The white exterior seems to increase its length, being unobscured by either trees or tall plantings. The central section of the house has three floors with a flat roof. Two lateral wings are disparate. The left extension is two stories high with an attic and an isosceles triangle-shaped, pitched roof. The right wing is two stories high with a roof slightly slanting anteriorly. A protruding canopy over the front door entrance is supported by two ionic columns resting on individual square-top pedestals four feet high.

Not ostentatious on the inside, embellishments are few. Attention is drawn to Darwin's study where he did most of his writing. One wall has floor-to-ceiling bookshelves. Most are vacant with some books sparsely placed. This room contains some pertinent memorabilia regarding Darwin's life and work. A round table four feet in diameter is conspicuous. On top of it are many specimens of seashells and stone artifacts that had been gathered during his explorations.

Close to this mini-museum table is an old globe of the world contained in a round, wooden, three-legged stand. This globe is a facsimile of nineteenth-century geography. The impressive reproduction has an embellished equatorial ring. A three-hundred-sixty degree median ring supporting the globe is accentuated by etched numerical degrees of latitude. At its northern axis, a brass time-dial caps the globe. When aligned with the prime meridian, the time can be determined anywhere in the world.

Chapter 6

Charles Darwin and Sigmund Freud

Science asks how things happen, then seeks to find out. Answers are sought by thinking, research, experimentation, and objective study. Religion asks why things happen and does not find all the answers even with maximum utilization of finite intelligence. Religion needed revelation by prophets and the vatic teachings of Christ and others to be understood. Each was related to the other for comprehension in continuity. Religious belief involves the gift of faith which facilitates and allows a transmission from the darkness of disbelief into the light of belief.

Mankind is brought into the knowledge of divinity by this gift of faith. God is met in Holy Scripture and the faithful touches Him in the holy sacraments. There are hints of God's presence less spiritually. These arise from scientific knowledge of the universe. The more profound the study of the world, the greater is its comprehension. Total encounter with the magnitude, complexity, natural harmony, and reality of all things created brings humanity to a closer security, appreciation, and proof of the existence of God. What has science to offer that is either similar to or can approach this bestowal of faith?

The accuracy of scientific observation is impeded because it depends on the trained intellect and depth of its utilization by the unbiased observer. Hence the ancient, pre-modern, and some modern scientists are deficient sometimes in their explanations. Therefore religion is minimized when it is not ignored completely.

Disinterest and denial of religion, whichever is more applicable, are the common ground upon which Charles Robert Darwin and Sigmund Freud (1856–1939), founder of psychoanalysis, stand. Inheritance of an acquired character was taught by the neurologist. Freud spoke, lectured, and wrote about inherited diseases—especially syphilis. He said to his patient, Albert Hirst, that his maternal grandfather's syphilis was a cause of his neurotic traits (Lynn 1997). Freudian teaching was that this disease was changeable as to dominance becoming stronger, not weakened, as it passed down from one generation to another.

When Hirst spoke of having tried to find comfort in alcohol, Freud told him not to worry about it. It would not help him, nor would it harm him, because he was Jewish. Alcohol was for the gentiles (Lynn 1997, 23). The psychoanalyst taught that resistance to a vice was hereditary and did not diminish in descendants. It was a survival trait not extinguishable because of ancestry.

Freud believed in the inheritance of an acquired characteristic which evolved in mutation from ancestors that survived and is augmented in offspring as an evolutionary biological process.* Freud admired Darwin "because Darwin also subscribed to the inheritance of acquired characteristics" (Lynn 1997, 91). Not to be too critical, it is not improper to question the expressions of both Darwin and Freud as to their scientific accuracy. Giving them the benefit of any doubt, they are credited with acting in honest conscience. Whether or not the conscience of each was educated by truth is questionable. If not, their dicta are a disservice to humanity. As scientists, they have not served well because science must be joined by a moral conscience trained in the differentiation between truth and falsehood. "It is my prayer that scientists will never forget that the cause of humanity is authentically served only if knowledge is joined to conscience" (Pope John Paul II, University of Padua, 1997).

Chapter 7

Darwin as a Scholar

Charles Darwin was a firm believer that expressions of emotion reveal the unity of human beings and their continuity with animals. In a portrait dated 1881, the year before he died, his facial expression is enigmatic. Perhaps this was because he was frequently ill with chronic diseases. Being unwell was the reason given for never leaving England again after his five-year voyage.

As a scholar he did more than write on the origin of species. He had published four volumes on barnacles, an obscure subject to merit a mini-encyclopedia. His research wandered into diverse biological realms ranging from worms to climbing vines, flowers, and animals.

In 1998 a new edition of *The Expression of the Emotions in Man and Animals* appeared. In it Darwin wondered whether human facial expressions are innate, the same in cultures around the world. And in support of his underlying theory that human beings are an extension of the animal continuum, he set out to show that animals have many of the same ways of physically expressing emotions as do human beings. The original book was published in 1872. The current edition is the first to include all the changes Darwin wished to make (Wiley 1998). Regarding human facial expressions, a medical study was made many decades ago at the Children's Hospital in Toronto, Canada. It concluded that the changes in the facial expressions of sick children were a valuable aid in the diagnosis of pediatric diseases.

A thesis presented on how a person can be judged by facial characteristics touched on Darwinism. Other facial tattoos are of evidential value. A small blue spot on the cheekbone used to be the reliable sign of an old Borstalian, until young men desired to look as though they had been to Borstal even when they had not, and tattooed themselves. "I suppose a Darwinian might consider this to be yet another example of imitation by natural selection, in which a harmless species of moth or butterfly comes to look exactly like a poisonous one. And it works, for such a person commands fear, respect or admiration" (Dalrymple 1999).

Darwin wrote many books in which he presented a vast amount of information. One of them, *Expressions of the Emotions* is an attempt to refute the contention that human beings were created separately from animals. His postulation is that human beings are the advanced products of the animal kingdom. Specifically Darwin was writing against a book by Sir Charles Bell* who considered the muscle in the human face that "knits the eyebrows" to be uniquely human. In the margin of Bell's book, Darwin wrote "Monkey here? . . . I have seen it well developed in monkeys . . . I suspect he never dissected [a] monkey" (Wiley 1998). Darwin's postulate of the kinship between monkey and mankind brought the monkey into the research laboratory as the favorite experimental animal of bioscientists.

Darwin's accumulated observation data suggested, but was unsaid, that animals are malleable and more predictable than assumed by previous generations of bioscientists. Pavlovian** conditioned reflex discovery confirmed his belief. This Russian revelation verified the physiological pliability in mammalian salivary secretions. Indeed, this was an exceptional contribution to animal and human gastrointestinal physiology.

A magnetic witness of evolution was demonstrated in the amusement world. Phineas Taylor Barnum (1810–1891), at age sixty-one, started his circus business with sideshows that mesmerized onlookers. Among the most fascinating of these was a humanoid called the missing link, which supposedly was the evolutionary connection between ape and man. This prosopopeia attraction was one of great fascination. Crowds came to view the missing link, which confused and/or conjured people into believing they were descended from monkeys. As a prestidigitator showman and circus producer, Phineas T. catapulted Barnum and Bailey Circus into the world of renown.

P. T. Barnum's missing link phenomenon received so much publicity that it broadened widely Darwin's fame. To many of the nonscientific community the spectacular exhibition was accepted as proof that men and women descended from apes. This belief continued even when Barnum's creature was unmasked as a fabrication. As a masterful innovator, Barnum popularized the use of exit signs for crowd-control. When visitors to the sideshows mulled in front of the various exhibits refusing to make room for other spectators, he had signs with large red letters placed over pathways. The words were "TO THE EGRESS." Thinking this indicated the way to an additional show, visitors went through a tunnel only to find themselves outside on the midway. Darwin indirectly assisted Barnum in establishing his circus reputation as a creator of amusing artifacts, bizarre stage constructions, and theatrical inventions.

The primary goal of Darwinism is to demonstrate that all human beings have certain innate qualities, including facial expressions. This, in turn, would be evidence of a common progenitor. His evidence was the sort that today would be dismissed as anecdotal. Yet Darwin compiled so much material from so many correspondents in so many different places that its sheer volume and variety became authoritative. In Australia, for example, as related in a biography by Adrian Desmond and James Moore, "missionaries and magistrates from Queensland to Victoria ceased converting and incarcerating to observe aboriginal ways" (Wiley 1998).

Credit is given to Darwin as an astute, keen observer of

everything. This included his children, his dogs, his cats, friends, and strangers without embarrassment wherever he was. Always mentally inquiring, Darwin questioned why human beings and animals did what they did. Then he correlated the similarity of their actions, seeking a common reason for the activity observed.

He used three principles as his guidelines. The first he called the principle of serviceable associated habits. By this he meant that certain actions could be of service in certain states of mind, and the same movements would be performed out of habit even when they had no use whatsoever. He offered pages of examples. A person describing a horrible sight will often close his/her eyes and even shake the head, as if to drive the sight away. Or a person trying to remember something, on the other hand, often raises the eyebrows, as if to see better (Wiley 1998). By expression Darwin included any bodily movement, posture, physical sign, plus facial expression. All these observations indicated the body language that was interpretable.

Darwin's second principle was termed antithesis. An incident included in this category is illustrative of it: A dog is prepared to attack a person, then recognizes him/her as its master, recoils, and then changes every aspect of its appearance. "None of the latter expressions are of any use to the dog, they are simply the antithesis of what had been before" (Wiley 1998).

The third Darwinian principle is based upon involuntary actions of the sympathetic/parasympathetic nervous systems. Among these listed by him are trembling and sudden expression of joy at a feat accomplished. Added to this is perspiration during stress. Present-day psychophysiologists of Darwinian orientation place an emphasis on heart-brain communication. It "is now the focus of contemporary research and theory on both emotion and health" (Ekman 1998).

Five reasons have been offered for the rejection of Darwin's book containing the explanation of the pertinent appropriateness of his three principles. These are presented according to Ekman's critique:

1. Darwin was convinced animals had emotions and expressed them. This theory was dismissed as anthropomorphism.

2. His data was anecdotal, not proven scientific fact that could be duplicated by other bioscientists.

3. As did many of his colleagues, Darwin believed acquired characteristics could be inherited.

4. Darwin shunned studiously the communicative value of expressions because he purposefully avoided the common idea of his generation that God had given human beings a special anatomicophysiological capability to form facial expressions.

5. The last reason touches on the current controversy over such ideas as sociobiology. In Darwin's day behaviorism ruled. People believed that human beings are products of the environment, and therefore that "equal opportunity would create men and women who were the same in all respects" (Ekman 1998). Most modern day bioscientists agree that we are creatures of nature as well as nurture. "Genetics, not culture, makes certain expressions universal" (Wiley 1998).

Seen by many intellectuals as a scholar *par excellence,* Darwin's thoughts propagate ideas in variegated scientific disciplines. Strange to relate, some are so bizarre that the source of origin is difficult to attribute to professors of high reputation, such as: creativity is laudable but its proposals should be credible. Perhaps incredulity is inheritable as a Darwinian paradigm.

Notwithstanding his reputation for erudition, Darwin failed to display it in his treatise entitled, *The Descent of Man and Selection in*

Relation to Sex. This book is divisible into two components. The first part demolished the belief that the universe was created for humankind. In the second half he presents observations of his earlier hypothesis regarding sexual selection (Darwin 1877).

Regardless of this fault, Darwin still stirs the imagination of bioscientists who fall back on his theory, intertwining it with their disciplines in the search for answers. The current faddism is in geriatric studies and research. Although the fountain of youth is not sought, researchers hope to find the basis for new therapies that could lessen the impact of aging on human lives. The search is not so much why slow aging has evolved, but to know how it is done (Holmes, Austad, and Ogburn 1998). It is evolutionarily worthwhile to devote intellectual efforts to slow the rate of aging (Austad 1998).

Chapter 8

Evolution and Rape

In a startling publication, an evolutionary biologist and an evolutionary anthropologist present an opinion on rape. It is proposed that rape "is, in its very essence, a sexual act [which] has evolved over millennia of human history" (Thornhill 2000; Palmer 2000). They are not in agreement on the specifics of rape's precise evolutionary function. One of the authors believes that rape has evolved as "one more way [for males] to gain access to females" (Thornhill 2000). In an explanation, the motivation is passed on by their genes to another generation. This is a sexual strategy for males who lack "looks, wealth or status" or see low costs in coercive copulation" (Thornhill 2000).

The other scientist, an evolutionary anthropologist, believes "that rape evolved not as a reproductive strategy in itself but merely as a side effect of other adaptations, such as the strong male sex drive and the male desire to mate with a variety of women" (Palmer 2000). Both agree that "rape has evolutionary—and thus genetic—origins" (Palmer). Rape is stated to be "a natural, biological phenomenon." The authors do not justify rape. "As scientists who would like to see rape eradicated—we sincerely hope that truth will prevail over the politically constructed notions about rape now in vogue—preventive measures must take into account rape's evolutionary roots" (Thornhill 2000; Palmer 2000).

A correlation between evolution and rape is a conjecture that is only speculative. No coordinated or coherent scientific proof removes it from the sphere of esoteric speculation. By its own indictment it remains an untenable surmise. Rape is an act by an immoral, multifaceted sociopath, in whom the mental aberration is of long duration.

Men who rape are unable to compete with women and are resentful toward them. Therefore they consider themselves inferior to them. Their experiences with women proved to them they are not equal to women and their male talents are inferior to those of women whom they have encountered. This is interpreted as a degradation of the person's manhood.

With hallucinatory irrationality, rapists are infuriated to anger because of the male/female disparity they interpret as true and real. Rape becomes vengeance as a demonstration of male distemper. Often, when attacking a woman, an assistant is needed. A threatening knife is placed under the chin with intentions to cut the victim's face if she resists. Women fear having residual facial scars. Men consider a cheek scar as a dueling sign of bravery.

The act of rape is a punishment of women who have lowered the self-esteem of the perpetrator. For the rapist it is restitution for the ill will he had to endure under humiliating circumstances. The act of rape is vindication that demonstrates male superiority over women and domination over them. This mental disorientation persists even when the victim is a woman unknown to the rapist. It is a generalized castigation of women who have chastised, denigrated, and maltreated him, women who are *in absentia* and not the actual victim but who symbolize those hated. Such is the natural history of rape according to interlinear evolutionists.

Chapter 9

Evolutionary Psychology

Into the discipline of general psychology has been insinuated the term *evolutionary psychology.* This means that "human impulses and behaviors are best understood as the products of natural selection" (Buss 2000). Accordingly, science can unravel many of love's mysteries and psychologically explain "everything from why we cooperate or cheat to why we find others attractive." These are the results of natural selection (Dugatkin 2000).

In surveys by evolutionary psychologists on infidelity, men choose sexual infidelity and women are less physical, selecting emotional relationships. Imaginative theorizing is being overcome in psychology by those who adopt the evolutionary theme.

Not all the proposals on behalf of this form of psychology have substantial proof. "Natural selection favors making an error in the less costly direction, hence the false positives" (Dugatkin 2000). Inaccuracies are found in this so-called new psychology because explanations given are based on scanty evidence offered as proof which may be groundless. Advanced in the publication, the concept of evolutionary psychology emphasizes infidelity, jealousy, and other emotionally misguided behavior.

The author posits that it is easier to explain male infidelity than when a women commits it. Believing reproductive potential is expressed when a man has many cohabitants, this is not applicable to women, whose reproductive capacity is restricted to once every nine months. Infidelity in women is a quest for superior genes, especially an exceptional son. Another, less complimentary excuse would be to establish a partnership with a man higher up on the social ladder than her current male companion or spouse.

Much of evolutionary psychology places women in an unsavory limelight filled with culpability. Often their infidelity may be well-planned. A woman may look for a man to take the place of her present mate, or she may hope for sexual gratification at a higher level than she has experienced. All these are theories but proof is absent.

Not satisfactorily clarified is the author's reference to jealousy. Without a preliminary discussion as to why this subject was chosen to bolster the concept of evolutionary psychology, jealousy is proposed as a positive force in this modern age. "Properly used jealousy can enrich relationships, spark passion, and amplify commitment . . . [An} absence of jealousy . . . portends emotional bankruptcy" (Buss 2000). In the evolution of desire there is a focus on jealousy which is described as "an adaptive emotion forged over millions of years" (Buss 2000). According to the professor, jealousy evolved over the years as a naturally selected beneficial emotion. It did arise from long-term love, not as a part of a capitalistic society, social status, character deficiency, neurosis, or psychosis.

Addressed to the clerisy is a reminder that any person who writes easily what he/she knows sometimes will write about what is not proven. For public acclaim there are many who will barter away the immediate jewel of their souls. Also, to not oppose error is to approve of it, and to not defend truth is to suppress it.

Chapter 10

The Concept of Evolution

Quite erroneously, the idea of evolution is attributed to Charles R. Darwin. Egyptians had an evolution theory two thousand years before he was born. Anaximander (c. 585 B.C.), by his own deductive reasoning, postulated that the long period of breast feeding during infancy for survival indicated that human beings developed from other animals who were nurtured the same way. Added to this belief was his idea that all life began in the sea. Change to newer forms of animal anatomy evolved by adaptation to the new environment to which they migrated. By observation Aristotle (c. 322 B.C.) arranges organisms in a developmental pattern from the simple to the more complex (Treatise on Animals) (Kitto 1951). Evolution does not connect them.

At the outposts of human knowledge on evolution, the first to project a mental concept thereon was Thales. Chief of the seven sages of Greece, little is known of him. By traditional hearsay he became well accepted as the father of science. Aristotle called Thales the man who made the first attempt to establish the physical beginning of humanity without any recourse to mythology. His friend was Anaximander, concerning whom history gives little information except to infer that he was a teacher on a narrow strip of land by the Aegean Sea. Anaximander's teaching thesis was the transmutation of species and that man as an entity developed from aquatic animals.

On this subject Aristotle was not an alien. How near he came to the general conception of evolution may be gathered from this excerpt: "Nature passes from lifeless objects to animals in such unbroken sequence, interposing between them beings which live and are not animals, that scarcely any difference seems to exist between two neighboring groups owing to their close proximity" (Robinson 1931).

The nineteenth century is known as the beginning of the modern age of biology. Much of this fame occurred after Charles Darwin returned from his journey on the *Beagle* and his explorations of the Galapagos Islands. The ship's captain was Robert Fitz-Roy, known as the father of the weather bureau. The captain lacked enthusiasm for Darwin. He was prepared to decline him passage. Ready to reject the volunteer naturalist, sailor-scientist Fitz-Roy had a change of mind when he realized the only reason for his dislike of Darwin was the shape of the future passenger's nose. On such a folly rested the decision for a momentous voyage upon which hinged an era in the history of biology.

If repudiated, Darwin would not have been away from his Shrewsbury garden for five years. Never would he have seen the naked savages of Ecuador, be amid South American fossils, and wallow with the fauna of the Galapagos. Remaining uncelebrated would have been the Malthusian-inspired words of Darwin: "that under these circumstances favorable variations would tend to be preserved, and unfavorable ones to be destroyed" (Robinson 1931). No doubt the scientific spirit was uplifted when Darwin composed and published his *Origin of Species*. He was not the first to originate the idea; however, after November 24, 1859, the day of his very successful publication, he had made his mark in the biocultural history of spectacular revolutionary theories in science.

When the radiant, splendid years of the renaissance had surfeited the intellectual gentry with artistic beauty, an interest in nature emerged. Inquisitive minds aroused speculations that questioned religion, the Bible, and Christian dogma. The great geographical discoveries and explorations in science gave rise to newer studies like geology. The finding of fossils led to more speculation on the origin of species.

Erasmus Darwin (1731–1802), the grandfather of Charles, proposed a theory of evolution fifteen years before his grandson's birth. It was in the literary form of a poem which had no status as literature and even less scientific merit (Waddington 1958). The French scientist, Lamark, in 1809 published a coherent theory that attempted to explain the mechanism of evolution. This publication appeared before Charles Darwin was born (Filice 1995).

Evolutionary biology is seductively metaphorical. Its evidence points in so many suggestive directions that its practitioners are naturally tempted to make global speculations. Charles Darwin understood this very well, and knew moreover that it could lead to unfounded ideas as well as innovative ones. Concerned with establishing his new theory as serious science, Darwin laid out a rigorous formula for evolutionary discourse, explicitly rejecting for himself and his followers the more speculative style of early evolutionists such as Jean Lamarck and his own grandfather (Reich 1997).

William James (1842–1910), the American philosopher and psychologist, inherited his transcendentalist literary psychology; however, he was forced to square its religious epistemology with the more rigorous scientific dictates of his academic ambiance. Hence he became a defender of consciousness as an efficacious force in the biological evolution of the species. As a medical student at Harvard he sided with the Darwinians (1860s). He began his literary career by writing favorably about the effects of natural selection on mental life. "Consciousness," he observed, "obeys the laws of variation and selection . . . Rationality and empirical dictates of the sensory world then select out what is adaptive and what is not. In this manner experience as a whole counts as a potent force in the preservation of the race" (Taylor 1992).

Restraint was not on the evolutionary agenda. Beginning with T. H. Huxley, evolutionary biologists arrived at a two-track solution to the problem. They published one set of books and articles to establish professional credentials, and a distinct but parallel set to appeal to popular audiences and to serve as an outlet for speculation. "This strategy has not been lost on mainstream biologists, many of whom see evolutionary biology as a field tainted by the imposition of cultural values. They pay lip service to it but in practice regard it as a less-than-orthodox subject for research" (Reich 1997).

Ruse argues that evolutionary studies have been shaped from the beginning by an overarching "concept of progress" that does not, despite its secular nature, fit comfortably into the scientific enterprise. In this methodical study, he tries to show how notions of social and moral betterment—and their perceived connection to biological progression from microorganism to man—have influenced the scientific thought of major Anglo-American figures from Herbert Russell Wallace to George Gaylord Simpson and Geoffrey Parker (Ruse 1997).

The case is not always convincing. Consider Ronald A. Fisher (1890–1962), whose achievement was to add nuance and mathematical structure to evolutionism by combining Darwin's theory of natural selection with Gregor Mendel's principles of genetics. Fisher was passionately interested in eugenics and believed, erroneously, that almost all human abilities are innate. Ruse asserts, but does not really prove, that Fisher's enthusiasm for human progress through breeding distorted his actual scientific work.

More compelling is Ruse's examination of the contemporary debate over Edward O. Wilson's theory of sociobiology, which posits that human social behavior can be understood in terms of evolutionary origins. Ruse makes the cogent point that while Wilson's enthusiasm for cultural progress has led to an explicitly stated belief in biological progress, the same enthusiasm in Stephen Jay Gould has led to a career built on energetic denial of biological progress. In this modern context, it does seem that evolutionary biology has become infused, polarized, and defined, by an underlying cultural value (Reich 1997).

As appraised, Darwinism is a theory of evolution by natural selection. It maintains that irreversible changes take place in the characteristics of organisms, occurring in successive generations related by descent. The theory that evolution accounts for the origin of all kinds of organisms now existing is opposed to the biblical writings on special creation, meaning that each kind of organism was created as such and therefore is not related by descent to any other species.

Natural selection is the principal mechanism of evolutionary change as originally suggested by Darwin in 1859. The theory that evolution has occurred by natural selection asserts that individuals varying in certain particular directions contribute more offspring to succeeding generations than those varying in other directions.

The theory of natural selection supposes that the enormous loss of potential offspring is not entirely random. Seemingly repetitious, the theory emphasizes that individuals varying in certain directions are more successful in leaving offspring than those varying in other directions. The successful individuals are selected by a natural process. Restated with other words, different variations have different "survival values" in the face of the hostile circumstances in which all organisms live. Of these variations, some are due to environment, some to genes, some to a combination of both. The word *variation* in its application is a general term for differences between individuals of the same species.

Only the gene differences are inherited. Thus only they can be of importance in evolution. If one variation has a higher survival value than others, it and the gene responsible for it must become more prevalent in the population in succeeding generations. Thus the less successful characteristics and genes will diminish proportionately. This selection of certain genes presents in a population for future survival is what is meant by natural selection.

In many populations, the main effect of natural selection simply may be to weed out certain mutants, or combination of genes, as they occur, thus maintaining the population fairly stabile in composition. But it is additionally the main agent of evolutionary change by incorporating into future generations of the population new genes or combination of genes with greater survival value as they appear by chance, or when an alteration of environmental conditions alters existing survival values (*Encyclopedia Britannica* 1996).

Like some other misconceptions proposed by evolutionists, the increase in size of descendants is equivalent to being better than their smaller-sized ancestors. Being bigger can be overrated. In 1871 the American paleontologist Edward Drinker Cope viewed the evolutionary history of horses and other species. Observing a pattern, he deduced that the longer a species evolves, the larger it gets. For one hundred twenty-five years Cope's rule held sway in scientific acceptance. Resolutely tested in 1996 by sturdy studies, it failed the test (Siemann et al. 1996). Bigger is not to be equated with better.

From large horses to small insects, the evolutionary theory has been downsized. Insect studies in the grasslands and savannas of Minnesota reported that insects thrive in an enormous range of sizes. Three ecologists collected 89,596 insects, representing 1,167 species and five orders. Studies of the samples taken gave a formulation to a new "power law" concerning insect populations. Namely, the number of species of a given size is proportional to the square root of the number of individuals within those species. For example, if one size category includes ten times as many individuals as a second category, the first category should also include 3.2 times as many species as the second one (Siemann et al. 1996).

If Cope's rule was correct, the expectation would be that the largest insects would be the most abundant. Siemann and his colleagues found that medium-size insects were not only the most abundant but also the most diverse. At the same time, peak sizes among the five insect orders varied more than a hundredfold (Santos 1997).

The Minnesota University study made Cope's rule quite

morbid. At the University of Chicago a paleontologist summarized a ten-year study of the evolutionary history of 191 mollusk lineages with the information that mollusks showed no overall tendency to increase in size. In fact the confirming conclusion was that just as many lineages shrank during that time (Jablonski 1997).

A fascinating aside on the theme of insects is not to be considered facetiously. In an experiment worthy of note, a focus on eyes was revealing. To learn more about how evolution produced the rich variety of animal life on earth, biologists took a gene for making eyes in squid and inserted it into fruit fly larvae, creating insects with eyes on their wings, legs and antennae. The bizarre experiment produced graphic new evidence challenging the long-held assumption that because there are so many different kinds of eyes in the world, these often highly complex organs must have evolved completely separately.

"For many years people saw eyes in different organisms and believed they must have evolved independently," said Stanislav I. Tomarev, a molecular biologist at the National Eye Institute who helped conduct the experiment. "The main implication of this experiment is that eyes in different animals have a common evolutionary origin," he concluded.

This work shows that a single gene found in such disparate creatures as fruit flies, mice, and squid plays a critical role in the growth of very different kinds of eyes in very different animals. "The initial step that triggers eye development is the same in very different groups of organisms," Tomarev said. "This was demonstrated before for flies and vertebrates. Now we bring another big group of organisms into the picture—mollusks. It's not just for flies and vertebrates. Probably, it's universal," was his opinion.

Other researchers said the studies were an important new piece of evidence suggesting that the first primitive eye may have evolved once, and then simply adapted into the various forms seen today to meet individual creatures' needs. But it is probably not enough to persuade everyone to abandon the long-held idea that eyes represented a quintessential example of divergent evolutionary biology. "Here may be too many unresolved questions to allow skeptics to buy into the bold hypothesis at present," wrote William A. Harris of the University of California in San Diego in a commentary accompanying the report in an issue of the *Proceedings of the National Academy of Sciences.* "But neither can they rule it out."

The hypothesis that the eyes of human beings, flies, and other animals evolved separately stood largely unchallenged until 1995. That is when Walter Gehring and colleagues at the University of Basel in Switzerland took a gene called "eyeless," which they believed was involved in eye formation in fruit flies, and inserted it into different cells in larval fruit flies, causing the formation of eyes on their wings and other parts of their bodies. That showed a single gene could trigger the growth of something as complex as a compound eye.

Gehring and his colleagues then showed that an equivalent gene found in mice, called Pax-6, could also cause fruit flies to grow fly eyes—strongly suggesting that eye formation shared "common developmental mechanisms," Harris wrote. In the new research, Tomarev and Gehring took the squid equivalent of Pax-6 and inserted it into various cells from fly larvae, again causing eyes to grow in various parts of their adult bodies. Although the eyes appeared fully formed, they are unlikely to be functional because they do not appear to be connected to the brain, Tomarev said.

The experiment shows that Pax-6 is either a "master regulator of eye development," or, at the very least, a "patterning gene" for eyes. Presumably, the gene interacts with other genes to carry out the intricate development of an eye, which is one of the most complex structures in nature. More research is needed to determine exactly what role Pax-6 plays in that elaborate cascade of events.

Regardless, the new research shows that eyes in many species have "a common evolutionary origin and are products of parallel rather than convergent evolution," Tomarev and his colleagues wrote (Stein 1997).

In a remarkable, intellectually provocative volume, two authors have undertaken the formidable task of seeking to give a coherent narrative on the historical development of what are now currently called the neurosciences. The writers place at the center of their account the postulate that the nervous system is the outcome of a long and involved evolutionary process. To them the modern human brain is the endpoint of a "bushy" continuum. They state it is equally important to grasp that within this process function has taken precedence over structure. The organism develops needs in the struggle for existence. Appropriate capacities emerge and new organs adapted to or are superimposed within the encephalon. Thus arose the functional and structural hierarchy that exists within the nervous system. These basic premises were established in the middle of the nineteenth century due to the work of three diverse Englishmen: the philosopher Herbert Spencer, the clinician John Hughlings Jackson, and the naturalist Charles Darwin (Marshall and Magoun 1998; Jacyne 1999).

Charles Darwin evaluated the expression of the emotions in human beings and animals (Darwin 1873). He examined the causes, physiological and psychological, of all the fundamental emotions in man and animals. The study was part of the evolutionary set-up. This was a contribution to the development of the theory of evolution. Dov Ospovat of the University of Nebraska in Lincoln wrote on Darwin's theory from the viewpoint of natural history, natural theology, and natural selection. The period studied was 1838 to 1859. His studies culminated in the publication of a textbook in 1981 with a second edition in 1995 (Ospovat 1995). It has become accepted generally as one of the most influential books about Darwin in recent years. Thus it is considered a prominent factual treatise among Darwinian scholars.

The evolutionary explosion ignited by clever biologists had a dichotomous appeal that was simultaneously sinister and sensational. To some scientists it was inspirational. Evolution was interpreted by a cadre of academic professors as philosophical naturalism. Humankind was the end-product of evolution. With this finality of development, mankind decides what is right and what is wrong. Hence there is no thought given to an educated conscience. Morality is not fixed by commandments, allowing men/women to change right and wrong according to circumstances. If evolution is a theory and not proven fact, it can be questioned and criticized like any other theory. As a theory there would be no objection to it being taught. When presented as biological fact, objections arise to it being taught.

From the Hebrew University in Jerusalem comes a comment from one of its fine scholars. Under the section "Letters from Readers" in the magazine *Science,* the following is excerpted (Santos 1997).

> *The oft-quoted statistic that the genomes of humans and chimpanzees are 98 percent similar, which Meredith F. Small also cites in her article "Family Values" [Science magazine, November/December 1997], still leaves room for a 2 percent difference. That is equivalent to 60 million nucleotide differences between human and chimpanzee. Or, assuming the average gene extends over about 10,000 base pairs (including upstream and downstream regulatory regions, exons and introns), the 2 percent difference in an average gene implies that there are some 200 nucleotide differences between its human and chimpanzee analogues. That comes to 14 million nucleotide differences distinguishing the 70,000 genes of humans from those of chimpanzees.The five million years, which I estimate to be 500,000 generations, that separate human and chimpanzee lineages have provided ample opportunity for founder effects, group, and individual selection, migration, and hybridization to have molded those differences at the DNA level into what Philip Lieberman, in his article "Peak Capacity" [appearing in the same issue as*

Ms. Small's article], calls ". . . a capacity for thought that had never existed before, and that has transformed the world." There is more than a "layer of veneer," as Ms. Small would have it, between humans and apes.

Progress in science usually comes about by slow starts with small steps that increase incrementally. Uncommonly, an occasional outburst of a great discovery results in genuine acclaim. Seminal scientific discoveries become major contributions after prolonged laborious research experimentation. Deep thinking brings forth new perspectives that yield unimagined results.

Darwin's assumptions had a volcanic beginning, only to dwindle into secular parochialism. But it has not evanesced into an infinitesimal ghost. Darwinism had little resistance in its assault against long-standing biological tenets. No fortified beam prevented the invisible battery that undermined theological teachings. Professing supposed originality it had appeal to the nonscientific and anti-religious mentality. Without strict adherence to the highest, optimal research norms, Darwin won disciples who either ignored or did not concern themselves with authentic research procedures as a means of seeking scientific truth.

No threat of disciplinary action was available to constrain him. Nothing required him to follow scientific standard procedures prior to publishing his theory. In turn the public arena of scientific colleagues did not discriminate between a wishful, theoretical novelty and proven facts sustained by scientifically validated postulates.

Darwinism gave empirical scrutiny of the human species a new impetus, if not a completely respectable *imprimatur.* But tracing mankind's development in children rather than in monkeys was an appealing enterprise. The child provided what Darwin's theory needed, an example of evolution in action. The spectacle of adaptive development was inspiring. At the same time, Darwin's theory offered what the new devotees of childhood needed: an example of systematic observation in action. In fact Darwin, as his

American followers eagerly emphasized, paved the way. "He kept notebooks about his own offspring, whose first tears, fears, reflexes, rages, noises, and subsequent social and verbal antics he tracked with a naturalist's curiosity and a father's empathy" (Hulbert Harris 1999).

Evolution is not a verifiable biological discovery. It seems to be an invented production emanating from a vivid imagination fired by ancestral influences kept alive by flickering memories of two paternal generations. The theory does not stand on firm underpinnings. Meticulous scrutiny and intellectual evaluation can identify its weaknesses.

Quality control, questioning, critical oversight, and demonstrable proof are the guiding assets of scientific endeavors. Adherence to these attributes has earned for science its special aura of dignified authority. To do less is a grievous fault. Not to challenge scientific assumptions is a disservice to the admired, idealistic scholarship of scientists. Also it can easily lead to bioethical obsolescence. With all the words written on evolution, one can conclude that the Darwins, *fils, père et grandpère,* believed that there is no conscious design in the universe. Therefore they accept no bare minimalism of creation.

Scientific conception stamps Darwinism as an indispensable enigma in bioscience. The theory of evolution by natural selection may be right. The science of genetics on which it is based does not aid its clarification. The disputes of the life sciences are solved by proximity to comprehension. The prodigious output of excellent popularizers of evolution has not facilitated reasons for logical conclusions. The weaknesses of the subject mean it is crammed full of argument.

A fundamental problem that has always faced the theory that species evolved by the operation of natural selection on genetic variation is that of explaining how so gradual a process could have produced such diversity without there being many identifiable cases of intermediate species. These are the links which are missing. Indeed, the shortcomings inspire one of Jones's most eloquent paragraphs, which is quoted as an example of the heights to which his style can rise:

> *Sixty billion people have lived since man appeared in modern form. To excavate every graveyard—and every fossil site—in the world would turn up no more than a minute fraction of their remains. The lost armies of the dead have a moral for evolution. They are a reminder that the geological record is a history of the world imperfectly kept, and written in a changing dialect; of this history we possess the last volume alone. Of this volume, only here and there a short chapter has been preserved; and of each page, only here and there are a few lines. However grand the monuments, and however firm the hope of eternal life, in the depths of time a Pharaoh in a pyramid has as little chance of immortality as does a soldier stamped into a bloody swamp. The history of ancient Egypt, of the present century—and of the existence of our own species—will soon be gone forever (Jones 1999).*

If evolution is correct, a greater abundance of promising examples of transition between two of the hundreds of thousands of species might be expected. The standard explanation, ably expounded by Jones, is that intermediate species must be squeezed out by the competition for the resources of overlapping habitats from both their predecessors and their successors. It does not seem that this explanation can be very easily reconciled with the claim that it was the greater "fitness" of the intermediates that led to their emergence in the first place. Somehow species must represent stabile resting points on the curve of adaptation.

Jones's chapter 8 centers on a presentation of the currently dominant basis for the taxonomy of species known as cladistics. The essence of cladistics is the concept of grouping by shared inheritance. Species are grouped in dependence on how many of their characteristics they derive from a common ancestor. This has the advantage of a kind of taxonomic democracy between the tiny minority of extant and the huge majority of extinct species. Thus, such a traditional Linnaean group of mammals as the insectivores, collected under the commonality of their grubby predilections, are broken up, with the hedgehogs splitting apart from the moles, shrews, and aardvarks. Cladistics, whose own evolutionary prospects are perhaps still *sub judice,* has certainly achieved an interesting reshuffle of the classificatory pack over scientific generations. Not only have familiar species been realigned, but, most strikingly of all, animals and plants themselves have been shown to be, in Jones's words, "a well-explored but minor part of the metropolis of life."

Can evolutionary biology be considered a genuinely predictive science? There has been no shortage of philosophers who have decried it as a kind of history rather than the source of a set of scientific facts which could in principle be projected into life systems yet to be discovered. Are we in a position to say how evolution must have gone in the past and will go if it continues? Jones cites the case of certain Caribbean lizards, adapted to an island with thick vegetation, which, on being experimentally moved to an island which had only twigs, duly evolved within a decade to match the characteristics of nearby species of similar habitat. It might be queried, however, whether this sort of experiment can justify the postulation of the emergence of entirely new species rather than of mere varieties.

At times, Jones is admirably detached from the human perspective. He is underwhelmed by the problem of biodiversity. Why should we worry that so many species are currently threatened by extinction, when so many, many more have become extinct in the past? In a limited ecological economy, evolution presupposes regular extinction. But we can only be relaxed about the fate of other contemporary species if we were indifferent to that of our own, which is intimately connected with them. And our concern should be all the keener for the facts, very ably expounded by Jones, that for most species artificial preservation means change and for all species change is irreversible" (Lawson-Tancred 1999).

The disciplined scientific imagination is hostile to the extension of Darwinian ideas into the study of man. Sociobiology and evolutionary psychology are no better than theoretical Darwinism. There is a persistent distance between the assertion that "behaviour comes from brains; and brains, like hearts or kidneys, are made by genes" and "the birth of Adam, whether real or metaphorical, marked the insertion into an animal body of a post-biological soul that leaves no fossils and needs no genes. It is to be hoped that Professor Jones will return to this topic when he updates *The Descent of Man*" (Lawson-Tancred 1999).

Chapter 11

Darwinian Protagonists

As the theory of evolution spread regionally, nationally, and world-wide, Darwin's reputation as a bioscientist increased remarkably and rapidly. Many outstanding biologists accepted his theory without hesitation, some with enthusiasm minus any effort to either question or confirm it. Obloquy from the bioscientific community was a rarity. Advocates saw in Darwinism the key to life in the word *survival.* This seems to border on Dianetics, whose first axiom is survival, as it posits the teaching that function monitors structure and the mind regulates the body and not that the body regulates the mind. Hence thought takes precedence over that which is physical.

Dianetics is an investigative, heuristic science built upon axioms. Workability rather than idealism is its motto. The claim made for these axioms is that by their usage certain definite and predictable results will evolve. Dianetics basically is a family of sciences. Its field of thought is classified into the knowable and the unknowable. One word is used as a guide. A standard is selected as the primary controlling axiom. The first axiom is to survive. Hence it is applicable as the lowest common denominator of the finite universe (Hubbard 1996). All forms of energy as units or species survive for some unknown end, for some unknown purpose, say the Dianeticists. In this manner of speaking Darwinism and Dianetics have found each other.

Preachers of Dianetics teach a second axiom which states that the purpose of the mind is to solve problems relating to survival. This number two axiom maintains the mind acts in obedience to a central basic command: survive. This is an additional connecting bond between Dianetics and evolution.

Studies in the field of evolution indicate that survival is the sole test of an organism. Whether the organism is treated in the form of daily activity or the total life of the species, every action of the organism is to lie within the field of survival. Acting in its environment it is guided by information received or acquired. Error and failure do not alter its basic impulse which motivates to survival. Sustaining the above axiom is another one. Briefly it affirms that the mind is the central direction system of the body, and resolves problems of survival because it directs or fails to direct their execution. Natural selection and other causes have established a primary rule for survival. "That the unit remain alive as long as possible as a unit and, by association and procreation, that the species remain alive as a species . . . [N]atural selection best preserves those species which follow this working rule. And the symbiotes of the successful species therefore have enhanced their opportunity for survival" (Hubbard 1996).

The creed of Dianetics is that all life is directed by one command and one command only—survive. Darwinism converts this exclamation into a verbal noun using the gerund, survival. Both share a common thought in this regard based upon the survival of the fittest. An additional tenancy joins them as each partakes of an equitable commonality that ignores dominical association and reliance. Contradictory as it may appear, the probability exists for the theory of evolution and Dianetics to advance, and at the same time remain idle and stagnate.

Pangenesis does not contribute to the sustenance nurturing evolution. This discredited hypothesis states every somatic cell generates self-representative hereditary materials that enter the bloodstream and eventually coalesce in reproductive cells, making possible the inheritance of acquired characteristics. Using the prefix *pan* indicates an involvement applying to all. Genesis is Greek for generation meaning the coming into being of anything. Synonyms are *origin* and *creation*. Darwinism is not supported by any application of pangenesis.

Chapter 12

The Stentorian Luminaries of Evolution

Human physiology indicates that mankind satisfies needs with the minimal expenditure of effort. Applicably Darwinian fundamentalists succeeded without any physical exertion. The time was propitious for the reception of evolution. On the day of publication, November 24, 1859, all 1,250 copies of the *Origin of Species* were sold. The world was waiting for Darwinism with eager anticipation. For one hundred years the idea of evolution was in the atmosphere. Darwin rushed into print to have priority over a rival, A. R. Wallace, who was promulgating a theory akin to Darwin's. Lacking some aspects, it was not quite a clone. Soon after publication the readers became infatuated with Darwinism (Roche 1987). This was due to the glowing review one month later by the appealing biologist, T. H. Huxley.

By his perseverance and steadfast efforts, Darwin owes in the greatest part the promotion of his theory to Huxley. He depicted with the utmost personal sincerity and disclosed frankness the transition through which his own mind had passed in establishing an adherence to a new sociobiological concept which was a breaking-away from the old tradition of creation.

In 1870 Shadworth Holloway Hodgson (1832–1912), an English philosopher, published a two-volume text entitled *The Theory of Practice.* This treatise articulated a theory that was termed epiphenomenalism. Hodgson hypothesized that feelings have no causal efficacy no matter their intensity. Mental states are present only as epiphenomena incapable of reflecting back to affect the nervous system (Wozniak 1992).

This view was taken up, popularized, and placed within an evolutionary framework by Thomas Henry Huxley (1825–1895). In an address in Belfast to the British Association for the Advancement of Science, Huxley presented one of the most widely circulated and cited influential lectures of the period. The subject was written under the heading "On the Hypothesis that Animals Are Automata and Its History" (1874). In this paper, Huxley suggested that states of consciousness are merely the effect of molecular changes in brain substance that has attained a prerequisite degree of organization. Therefore animals are conscious automata.

The thrill of international fame stirred him on to a great glory he no longer had to just dream about. He enjoyed this living experience contemporaneously and in future years of accolades. Huxley was the eponym of commingled sentiments of awe, rebellion, and admiration that brought him much reverence. He was not a man of fastidious reticence. His public impact was large in propagandizing evolution. He was the single most distinguished contributor to the advancement of the Darwinian theory. Speculation was not raised as to Huxley's motivated enthusiasm. He did not appear to be an unscrupulous man preying feral-like on his foes. As a scientific apostle he never eschewed a challenging oral conflict. His enthusiasm was unquenchable even when overturning the conventional ideas of basic biology in favor of a cryptogenetic doctrine of which absolute proof is lacking. Assertion in behalf of a theory not sustained is an ineffective demonstration of its future verity.

With an eminent ability for persuasion, Huxley achieved a prominent position in the scientific fraternity. He deserves remembrance as making vivid a biological theory to those persons of admirable qualities founded upon long moral and intellectual cultivation. None of his major recruits had to feign their lofty credentials. They possessed them in the fornix of their erudition. If his presentation on the origin of species was traceable by electroencephalography, luxuriant brain waves would be recorded. Huxley used words to convey their literal meaning. Expressions and utterances were selected deliberately to emphasize their intended understanding.

Huxley's lectures always conveyed authoritarianism. While he spoke, his audiences could not exclaim in unison that his voice glowed with simplicity, purity, serenity, kindness, courtesy, moderation, modesty, or delicacy. These qualities were never inchoate to his personality. No one is thanked for qualities that are absent. Although pride was evidenced in his demeanor, neither snobbery nor prudery maligned his name. He never gave the impression of surliness even at the height of acrimonious debate.

Always he demonstrated uniform enthusiasm for the subject presented and the subsequent discussion. Huxley knew how to speak well, having mastered the art of public speaking at an early age. No necromancy was needed to mesmerize his listeners. His deep and extensive learned scholarship was evident on all occasions. Eventually he was sought after as the most competent source of information on evolution and become a living, implacable monument to it.

Eventual acceptance of the theory of evolution was due in no small measure to Thomas Huxley, as the theory's strongest supporter and interpreter. While convinced of the essential validity of the theory, he nevertheless continually questioned various facets of it. Huxley was the first to apply evolution to paleontology (Huxley 1892).

Among his terse epithets that have been repeated without caustic reaction is this quotation. "Social progress means a checking of the cosmic process at every step and the substitution for it of another, which may be called the ethical process, the end of which is not the survival of those who may happen to be fittest . . . but of those who are ethically the best . . . I do not see how selection could be practiced without a serious weakening. It may be the destruction of the bonds which hold society together" (Huxley 1893).

Without contradicting negativism, during his professional lifetime Huxley was looked upon as one of the most eloquent of English writers on science. As a foremost British biologist, his advocacy of Darwinism was bold and pretentious. Strange then to relate, in two prestigious publications, one on animal classification and the other on physiology, evolution received no major recognition by him (Huxley 1869 and 1879).

Darwin's luminary champion, T. H. Huxley, had an interest in fishery. He served on three British fishing commissions. His classical terse statements on fish were frequently quoted but did not elevate him to the stature of an aphorist. He "argued that herring (and by extension, cod) could never be fished out—nature, in the Victorian view, being indestructible. Cod do find lots to eat, swimming with their huge mouths open ingesting whatever goes in the massive orifice" (Kurlansky 1998).

As to destructibility, there is a dearth of cod off Newfoundland and the coast of Ireland due to fishermen and fish-eaters who are phenomenally proficient predators. Contributors to the decreasing cod population are Englishmen who crave fish and chips. Additionally, Basques make a dent in the cod kingdom by their relish for a favorite dish called *bacalao a la Vizcaina*. Another reason for the depletion of cod is the pediatric need for cod-liver oil as a source of fat-soluble vitamins. New Englanders hankering for cod chowder (and clam chowder) made it a favorite dish, popularized by Daniel Webster, whose oration on cod chowder is among the recorded treasures of the United States Senate (Kurlansky 1998).

Huxley's theory of indestructibility seems quite weak. His knowledge of fish failed to include that the cod "species is stable only if each female, in her lifetime, produces at least two offspring that survive" (Kurlansky). For preservation, the rivalry race between reproduction of the species and the masterful efficiency of catching cod must be imbalanced by Darwinian natural selection. This may be impossible, just like Huxley's adage of herring/cod indestructibility is false. This maxim marred his scientific sunniness and made his hagiography less redoubtable. So his glory was not a radiant effete although his fame remains. Also Huxley's knowledge of fish was not within the ambit of his supposed encyclopedic intelligence.

Anthropologists studying ancient times in Africa have enlisted the aid of geneticists. Modern DNA is providing information to a supposed division in the ancestral lineage of human forebears. Analysis of a minuscule mutation in a single gene that took place two hundred thousand years ago is the basis for the speculation. The approximate estimation predates human beings by seventy thousand years. These researchers contend that the ancestors of Africans and non-Africans separated into two divided, distinct populations (Hey, geneticist, and Harris, anthropologist, 1999).

The conclusion reached was founded upon the study of a gene termed PDHA in sixteen Africans, nineteen non-Africans, and two chimpanzees. These studies, caution the discoverers, do not imply that the two populations evolved into modern human beings independent of one another. But they do maintain that the separated groups probably intermingled, allowing genes to flow between them. "Beneficial genes would have been favored by natural selection, and ultimately the two populations would end up virtually identical" (Hey, Harris 1999). So much for this published finding as a support for evolution.

While this study is provocative, other researchers of human evolution would like more proof. Genetic data rely on broad assumptions about time scales and are thus subject to large margins of error. "A geneticist at Radcliffe Hospital in Oxford, England, hails the work as unusual but adds that the estimate of when the split occurred could easily be off by 100,000 years" (Koerner 1999).

Another pro-Darwinian evolution event in the United Kingdom has flamed current enthusiasm for the never fading theory. "The project to grow a human heart in a laboratory is the latest victory for Darwinian fundamentalism, the new religion which has replaced Marxism in academia" (Johnson 1998). Experiments with human tissue and genes, which hitherto were prohibited areas for bioscientists, are now rapidly acquiring respectability as the Darwin cult takes over bioethics. *New Scientist* reported that "until a year or two ago, the taboo on human germ-line engineering was absolute. But now most experts say they'd be surprised if designer babies are not toddling around within the next 20 years or so" (Johnson 1998).

Human germ-line engineering—the altering of a human sperm or egg to effect changes which can be passed on from one generation to another—is unlawful in Britain and in twenty-two other countries that signed the Council of Europe convention which bans it. In America, it is not yet authorized by the U.S. Food and Drug Administration. But the rapidity with which technology is becoming available and the rise in public opinion toward babies made genetically immune to most chronic diseases seem to ensure its legalization. Once America takes the road to biological Utopia, the world will follow. The rise in Darwinian fundamentalism supplies the spiritual background to make this new adventurism seem not only inevitable but ethical, and morally correct.

One has to admit G. K. Chesterton was right in his statement: "When men cease to believe in God, they will not believe in nothing, they will believe in anything." There is in the human make-up a need to focus on a belief. When God was denied, especially among the intelligentsia, they replaced Him with someone else. Many in politics and the social sciences accepted Karl Marx. Others in biology followed Darwin. It mattered not that Marx was intellectually contemptible. A man who never did any research, who faked the evidence, plagiarized the ideas of others, and whose work was mainly put together from chaotic notes by his follower, Engels. Such blemishes did not disturb Marxist believers any more than the behavior of Yahweh in the Old Testament put off Jews and

Christians. Marxism was the worship of abstract ideas. Marx was a personal deity to the point where Stalin accorded him quasi-divinity and Mao Tse-tung always spoke of death as "going to join Marx." By the time the Marxist religion collapsed in irretrievable ruin at the end of the 1980s it had claimed over 100 million sacrificial victims on its altars, and in recidivist China these hecatombs continue (Johnson 1998).

It was predictable that a new hero would take Marx's place among those in academia. Marx was not like Darwin, who was a genuine scientist. On a distinct occasion Darwin refused Marx's invitation to strike a Faustian bargain. Darwin was modest in his claims. He was reluctant to have his work used for ideological purposes. "But it always has been: first by the anti-religious campaigner T. H. Huxley, then by the social Darwinists, then by both the Nazis and the Stalinists, and now—more directly—by the scientific triumphalists" (Johnson 1998). Biology has superseded physics as the fashionable science on the campus of universities and in the media. Inevitably, this has led Darwin to displace Newton and Einstein in the egghead pantheon. Moreover, biology now has its own technology assisted by engineers which is expanding and proliferating at breathtaking speed. Now it is emitting intoxicating bioethical fumes which provide the new opium that obtunds the intelligentsia. This new technology is increasingly financed, not, as was chemistry and physics, by governments looking to the distant future, but by ordinary, anxious people looking for better bodies, longer lives, and healthier children in the immediate present (Johnson 1998).

Marx summed up his own historic significance when he wrote, "The philosophers have only interpreted the world in various ways. The point is to change it." For most of the twentieth century Marxists got the chance to change the world. One hundred million murdered souls plus the ruins of present-day Russia and its satellite nations testify to their efforts. Darwin not merely described how species evolved but also, implicitly, indicated how this evolution could be accelerated, changed, and directed artificially. The new technology is now here. Bioscientists (with big business not far behind them) are avid to use it. Darwinian fundamentalism provides the ethical justification for the bioengineering that at times defies nature. The twenty-first century is building up for a greater moral catastrophe than the preceding one (Johnson 1998).

Dr. William C. Minor, a volunteer editorial contributor to the formation of the *Oxford English Dictionary,* was for a while a devoted reader of T. H. Huxley who is listed as a major Victorian biologist and philosopher. He is credited with having coined the term *agnostic.* Minor reflected similar feelings with added negativism. Both shared the belief that since the laws of nature, as they interpreted them, could quite satisfactorily explain all natural phenomenon, they could not find any logical need for the existence of God (Winchester 1998).

Chapter 13

Darwinian Antagonists

By its ever increasing knowledge, science has had many triumphs in the twentieth century. The benefits bestowed by magnificent discoveries, inventions, and creativity have improved human welfare. With all the beneficial productivity accredited to it, science has no immunity against vitriolic criticism. Proud achievements are insufficient to overshadow the mistrust and suspicion that are thrust against scientists.

In the academic arena there are antagonists of science who allege fraud in experimentation, plagiarism, false documentation, fudging, massaging data, and other misrepresentations. Václav Havel, the widely admired Czech writer and statesman, has vigorously expressed his disenchantment with the ethos of science: "Modern rationalism and modern science . . . now systematically leave [the natural world] behind, deny it, degrade, and defame it—and, of course, at the same time colonize it" (Bishop 1995).

Even influential scholars not active in the discipline of a science have joined the band of critics. Representative George E. Brown Jr., a Democrat from California who is a trained physicist, has stated that his faith in science has been shaken—if not diminished. He suggests that Congress and the "consumers" of scientific research "may have to take more of a hand in determining how science is conducted, and in what research is funded" (Bishop 1995).

An English professor calls attention to the biological inferences in the writings of F. Scott Fitzgerald. The author of *The Great*

Gatsby wrote before the Scopes trial that Victorians "shuddered when they found what Mr. Darwin was about" (Bender 1998). Fitzgerald tacitly joined the comedians who ridiculed in their fashions those conservative persons who did not accept their animal derivation. In an early writing, F. Scott describes a character as "a hairless ape with two dozen tricks" (Bender 1998).

Few can guess the extent to which Fitzgerald's interest in evolutionary biology shaped his literature. He was particularly concerned with three interrelated biological problems: (1) the question of eugenics as a possible solution to civilization's many ills, (2) the linked principle of accident and heredity (as he understood these through the lens of Ernest Haeckel's biogenetic law), and (3) the revolutionary theory of sexual selection that Darwin had presented in *The Descent of Man and Selection in Relation to Sex* (1871). "The principles of eugenics, accidental heredity, and sexual selection flow together as the prevailing undercurrent in most of Fitzgerald's work before and after *The Great Gatsby,* producing more anxiety than love from the tangled courtships of characters he deemed both beautiful and damned" (*Woodrow Wilson Quarterly* 1999).

Has Darwinism elevated or diminished the respect one has for science and researchers? Ambiguity is applicable to the theory of evolution. With it arises moral conflict. Choice may result in being branded adversely as a bioscientist by those who have an ever-increasing power in the universal field of scholarship. Pensiveness arises when it is realized that more than a century after the publication of Charles Darwin's *The Origin of Species* (1859), and seventy plus years after the Scopes trial in Tennessee dramatized the issue, the same controversy is being fought without predictability.

Surveys of adult Americans indicate that only a minority accepts evolution as an explanation for the origin of the human species (Bishop 1995). No successful method has been found which can give a logical comprehension to evolution. It seems very few laymen are endowed to understand it. Even for educated members of the public, evolution is largely a mystery. For evolutionists the laboratory is not their workplace. Somewhere they must have a hiding place where discussions take place. Until clarification is made that eliminates doubt, scientific bewilderment prevails and evolution remains a theoretical proposition without inflexible substance.

Human beings are the *summum bonum* of creation. This is not denied by evolutionists whose theory stops there as the end-product in the kingdom of life. How humanity arrived at its excellence is the point of dispute. It is the intellectual departure among those who disagree with Darwinism. A recent postulated opinion states: "The human species has evolved by natural selection in the developmental biases of mind and behavior, hence in human nature, just as it has in the anatomy and physiology of the body" (Wilson in *Wilson Quarterly* 1998). Nature defines each species binding individuals of each together without any crossing over of one species into the other.

In 1891 the Finnish anthropologist Edward A. Westermarck reported his thought that propagated into a major psychological addition to the discipline of anthropology. More contemporary research has refined the rule with an anecdotal hypothesis. "When a boy and girl are brought together before one or the other is 30 months of age, and then the pair are raised in proximity, they are later devoid of sexual interest in each other; indeed the very thought of it arouses aversion" (Wilson in *Wilson Quarterly* 1998). This fortifies the rational understanding of the consequence of inbreeding and the avoidance of marriage of those allied by the propinquity of consanguinity.

Today's science appears to allow outrageous possibilities. This encourages fiction writers and redactors of screenplays to have flights of unbridled imagination. Opposite to such thoughts are the physicists who predict faster-than-light travel across the galaxy and learning to make new universes to prescribed specification. Relevantly the new history of the universe is summed up thusly: "Hydrogen is a light, odorless gas, which, given enough time, turns into people.* When the universe began, it consisted mostly of hydrogen. That gas condensed into galaxies of stars, in whose cores heat and pressure fused atoms into heavier elements, including those necessary for life. Some of those stars exploded, spewing the heavier elements out into space. New stars and planets formed, including our own. On one of those planets, life appeared . . . None of this could have happened unless all the physical constants (the speed of light, the charge and mass of the electron, and similar numbers) were just right" (Wiley 1995).

This comment and notations on the universe do not sustain the theory of evolution. Speculation on the heavier elements necessary for life argues that the formation of life as proposed would signify a fully developed entity with a fixed substantial form that may grow but was incapable of metamorphosis. Hypotheses are malleable, often used to sustain that which is doubted and highly controversial. Worthy of intellectual discussion but difficult to accept is this phenomenon on the big explosion that brought about the universe. Who created the explosion and the hydrogen?

Peril of misinformation is not alien to science. Danger lurks when it contributes to the destruction of educational productivity. This results when accommodations are made to assist in bolstering transient proposals, appealing theories, popular hypotheses, and unsubstantiated opinions. The same evil applies when a scientist or someone else adheres to the philosophy of publish or perish in order to be in the limelight, however brief it may be.

Such actions encourage fraud, deceit, and deception. They are a departure from ethical conduct and deny syllogistic reason as an integer of scholarly discourse. Lack of dignified mental behavior is a blockage against fertile intellectual stimuli. Hence recognizing real facts becomes difficult. Scrutinizing participation in the interpretation of mental thoughts may become impaired.

As a consequence, incompatibility foments between fact and speculation. Such intransigencies can promote a never-ending search for fake reality which begets furtive prejudices and parologisms.* A pseudo-reality that has no intrinsic essence is assumed to be correct. The propulsion is zany academic interpretations that are either preconceived or conjectured. Surreptitiously and insidiously whimsicality enters scientific endeavors, research, teachings, and publications. All contribute to conclusions that are either untrue or camouflaged under terms that are neither accurate nor specific. Such distortion proposed as knowledge is solecism.* A countervailing capability to create new ideas is lost. When misguidance is produced by anger, envy, excessive criticism, and dispute, the brain can conjure up far-fetched thoughts. Often that which is produced mentally is difficult to believe. From controversy frequently comes exaggeration and exaggeration does the work of falsehood, unaware of the damage that has been done and continues unthwarted. So too, it is conducive to speak and write disparagingly of the best of things.

According to Darwin's theory, and hence supposedly the foundation of all biological and behavioral thought, learned responses are not inherited. Rather, all life is said to be directed by chance, by a dumb roll of genetic dice as it were. Thus, for example, the ancestral bird develops wings purely as a biochemical function and not according to some inherent thrust toward survival. Yet the moment survival is introduced as a pervasive drive, passed on from cell to cell, we are introducing an intelligence behind the scheme of life—an (unknown) X-factor (Hubbard 1996). Initially termed by Hubbard, the X-factor shapes and gives meaning to life in ways that Darwin simply could not explain. All life is directed by one command only—survive (L. Ron Hubbard 1996). Darwinism is not dominical. It is neither associated with nor does it have any reliance on the deity.

Survival of the fittest presupposes that those who survive become progressive strangers with each new generation. If this postulate is true then diseases should be less potent against the improved immunity of the latest generation. Unfortunately this is not totally correct. Although each generation lives longer than the previous one, longevity in itself does not eradicate major illnesses.

Young people die from virulent infections and disease. Newer diseases have afflicted all age groups with disastrous results. Pathology does not affirm evolution.

Apes are people-friendly, which may be interpreted as being of the same lineage by Darwinian proponents. Without valuable, authenticated research on the descent of mankind from monkeys, there are significant limits to this theoretical assumption.

A contradictory scientific fact can be used to cast serious doubt on this contention. "For example, early research in monkeys indicated that the rubella vaccine did not cross the placenta. Yet subsequent research showed otherwise, rendering the vaccine unsafe for human use. Subtle differences like these can have dramatic effects on the safety of drugs or procedures when findings based on animal models are applied to human beings" (Green 1998). This immunological study is recent information arising from biomedical research that does not support evolution as a scientific frontier. Predictably, additional research of this caliber fosters a bleak prognosis for Darwinism even if the controversy does not abate.

Darwinian evolution produced much smoke from the fire of criticism that contested the morality of its proposition. Unlike actual smoke, the theory did not dissipate with the wind-blow. Although the rhetoric was continual, forceful, and voluminous, it was ineffectual in dissipating the clouding, biological smoke. Words merely turned it in different directions.

Like a train locomotive whose furnace-fire produces smoke as the machine goes onward, leaving the accumulated smoke behind, the unconventional theory of evolution continued to move forward. With all the propulsion given to evolution by its supporters, a kernel of wisdom was missing which was a detriment to it. No proven truth for its veracity was available. Thus it lacks the luster to crown it the intellectual scientific jewel, the glory, and the outstanding good fortune of the nineteenth century.

Critics parsed Darwin's publications and speeches. They did

not conclude that his theory was to be looked upon as gaudy superficiality which relegated him to a confined, circumscribed, scientific demimonde. By the same evaluation no one found a basis upon which to construct a Darwinian logos to augment an established cosmic reason to confirm an orderliness for evolution in the biological world.

On a more symbolic level, inherent defects in his theory denies Darwin the major role in the nineteenth century's majestic drama of bioscientific accomplishments. Admiring colleagues and audient neophytes did not cease their adulations of him, in spite of the dappled splendor of his theory's appeal.

Regarding mental diseases, Darwin is silent. Although the physically fit are visible the mentally ill have not diminished. Contrarily, psychoneuroses are with us still. Newer diagnoses have been arrived at which add to the population of those with mental disturbances. The voluminous number of patients has prompted the elimination of older treatments as ineffectual. Since 1940 mental therapy has become a lucrative pharmaceutical matter.

Evolution is not an infallibility doctrine that is an integral nexus in the vast logical network of scientific thought. Nonetheless one should be chary of impugning the intelligence of those who believe in its utility under certain circumstances. Ignorance and hostility should not block out this theory as a callow inappropriateness that disallows religion.

Although constructive criticism can be beneficial, destructive criticism is not to be categorized as mental immaturity. It is a requirement that those who assume the critic's role must not be unfledged, but be knowledgeable of and at least understand the primordial premise of the matter discussed. Deep-seated prejudice and premeditated negativity steer the critic away from the obligation of performing a task with fairness, impartiality, sincerity, and honesty. These deficiencies diminish any beneficial value of criticism even as it is irresponsible, artificial, and deceptive. Also, excessive praise may be criticism in disguise.

The tenets of evolution may have some validity but scientific proof must be a condign fact to sustain its acceptance. When particular proof of one dictum is approved, it does not grant automatically a license to science to proclaim that all of it is true. Such fault is analogous to Einstein's relativity which has merits in the realm of physics. When relativity was applied to philosophy as a universal premise indicating that all things are relative, contrary opinions were many.

So it is with Darwinism. Fallaciousness arises, casting doubt when specific claims are enlarged to encompass more than was intended. Propaedeutical, apodictic facts are a *sine qua non* to buoy up the theory of evolution. The underpinnings of evolution are not self-evident even as they have not been clarified with exactitude. Hence it seems that science, when applicable to Darwinian evolution, propels one to believe it argues for the end of logic and the beginning of a new hypothetical cosmology and speculative anthropology.

Chapter 14

Mendel and Darwin

Against the nineteenth-century popularity of Darwinism was a contrarian view on its biological usefulness. After 1900, there was much speculation on Gregor Mendel's attitude toward the evolutionary theory of Darwin. Some long contradictory conclusions had been drawn. Some said Mendel rejected Darwin's theory, while others claimed he had supported it. The former based their claims on a biographical notice on Mendel published by Bateson (1902) which states: "With the views which were at that time coming into prominence, Mendel did not find himself in full agreement" (Coleman 1967). At that time Bateson opposed the Darwinian theory of continuous variation, and in his criticism he made use of only half a letter from Mendel's nephew, Ferdinand Schindler, written in 1902, where he states: "He [Mendel] read with great interest Darwin's work in German translation, and admired his genius, though he did not agree with all the principles of this immortal natural philosopher. He often told us nephews that we would find in his bequest papers for publication which he could not do during his lifetime. But we did not get anything from the monastery" (Coleman 1967).

Writing many years afterward, Mendel's nephew probably exaggerated his uncle's opinion of evolution. In the same year that Ferdinand wrote this quoted letter, his brother, Alois, published a speech given in Mendel's native village, in which he stated that Darwin's theory had aroused interest among both scientists and theologians. While many scientists merely launched into further daring theories, Mendel took the only correct way, which was to examine the theory through experimentation (Krizenecky quoted by Orel 1965, 8).

The two Schindler brothers studied medicine in Vienna. They were equipped by education to understand Darwin's theory even while Mendel was alive. But it was not until many years later that Ferdinand recalled a manuscript of Mendel's relating to Darwin. His brother, Alois, did not mention it. According to Mendel's biographer, the above is verified (Iltis 1966, 103). When Darwin's name came up in a discussion, Mendel said that "the theory was inadequate, that something was lacking" (Iltis, quoted by Orel 1966).

Later in his life Mendel was referred to as an opponent of the Darwinian theory for ideological reasons. Unpublished but understandable, Mendel was aware that Darwin was the herald of antagonism to religion in science. As a substantiating example in 1933 an excerpt from a book is quoted: "Into the important but dark realm of causes and laws according to which changes follow, about which he knew nothing at all, his contemporary, the Bruno* hermit, Gregor Mendel (1822–1884), introduced the first light through his brilliant discovery of the laws of heredity, which has practically the same import as Copernicus' laws, and brought to an end the scientific catastrophe of Darwinism" (Richter 1941).

"Mendel offered the most convincing evidence as the most important witness for the prosecution against the principles of Darwin's theory" (Matowskova and Matousek 1959). Mendel defended the constancy of species: "Lysenko and his disciples also claimed Mendel to have been opposed to Darwin's theory of evolution, without putting forward any evidence" (Hercik, Novak 1952).

Chapter 15

Sociological Changes

Marxism as an intellectual critique of society was a deconstruction of it. Simultaneously, it was a rewriting of history to reflect racial and other civil interests of various segments of national populations. Correspondingly, Marx's philosophy was the propulsion of a so-called scientific socialism. Epistemologically it emphasized the primacy of science's thought mechanisms, whose method and content linked the knower and the knowledge known to exist in an objectivity consistent with a relationship to reality.

As a similitude Darwin was redoing creation, correcting physiology, and altering biological functionalism. What Karl Marx found admirable in science, its precise thinking process, Darwin dismissed. Advocating consistency with reality, Marxists could not be staunch, steadfast, faithful supporters of evolution.

Little has been written on Nazism and Darwinism. What is available is uncomplimentary to the theory of evolution. Nazi Germany was a biocracy. As a government it was not admirable. Fundamentally, it was implemented by policies predicated on a distorted, perverted, nationalistic pseudo-science. "From the Darwinian concept of evolution came the idea that man could influence this process by selective breeding" (Proctor 1988). "Out of evolution emerged the notions of racial hygiene termed *rassenhygiene* and scientific racism" (Lifton 1986; Mitscherlich et al. 1949).

"National Socialism saw itself as a biocracy, that is, a polity with the power to heal the society of its diseases and restore it to full health. Hitler becomes the master physician, operating on the body politic with the help of German doctors. The second metaphor compares some citizens to diseased organs rather than autonomous members. Thus it becomes a matter of duty to extirpate the offending parts (Gaudiani 1996). Nazism wanted a pure race, a survival of the fittest. In order to obtain this goal, murder was not omitted. Elimination of undesirable racial elements was the rapid method of applied biology. Evolution had the same concept but depended upon natural selection to eliminate the unfit. Although there is no proof, many physicians are of the opinion that Darwinian evolution influenced the National Socialism of Adolf Hitler (Panush 1996). The sought after end-result was identical. The methodology was not the same. Nazism was to achieve it by radical revolution. Darwinism's postulate was to be the result of evolution over generations.

Among those special properties distinguishing human beings from other animate creatures are language and the brain. Questions pertinent to them are many, especially a scientific query as to which came first. Typical conclusions center about the belief that the brain developed first. Some social linguistic experts believe that voice capabilities are inchoate to the brain, pre-programmed from conception, and not learned by rote (Deacon 1997). A missing gap is needed to answer human origin from a lower species in this regard.

A powerful thought has been written that does not add positiveness to evolution. Not indulging in an *argumentum ad hominem*[*], a biological anthropologist has written that the growth of the brain did not precede and permit the evolution of language, but that the brain and language co-evolved. "An increase in the size of the brain's prefrontal cortex permitted the development of symbolic capacity. [E]ven as the complex linguistic structures deriving from that capacity, selected in the Darwinian sense, a brain had to have a large enough prefrontal cortex to make use of this novel form of communication" (Lakoff 1997). "The major structural and functional innovations that make human brains capable of unprecedented mental feats evolved in response to the use of something as abstract and virtual as the power of words" (Deacon 1997). Some primates have large vocabularies and relatively abstract ways of expressing information. On the other hand, human linguistic communication has taken a giant leap beyond any nonhuman system in that it uses symbolic reference. Our rules of grammar are symbolic, as are the categories (nouns, etc.) that they incorporate. "The vocabularies of all human languages encode abstract ideas. Mere increases in the numbers of words available in a primate communicative system, or longer strings of words, would not make them equivalent to human languages" (Lakoff 1997). According to some authorities, the language and linguistic presence in human beings is not a support of the evolution theory. A vast spatial void exists in human conversation and voice expressions in lower species.

Chapter 16

Darwinism and the Law

Coincident with the establishment of pro-evolution scientific groups, with Huxley as the gauge, was the emergence of strong biomedical moralism critical of Darwinism. A psychology of difference was provocative of long-delayed introduction to the dispute between religious scientists and those inclined to either agnosticism or atheism. The opponents of evolution demonstrated vast knowledge with intellectual capacity to express it in their criticism of Darwinism. Historians of science have mentioned scientific research performed for courts of law in behalf of creation. A revolution in legal advocacy occurred when scientists were called to testify in the Scopes trial in Tennessee concerning the teaching of evolution. This was the outstanding trial of 1925 taking place not only in Dayton, Tennessee, but in the entire nation.

Trial attorney Clarence Darrow (1857–1938) with William Jennings Bryan (1860–1925) brought luster to the arguments for and against evolution. Many turned away from the theory, which was presented as a *fait accompli* when there was no documented corroboration that was verifiable. The strongest, most vociferous opponents to the evolution speculation diagnosed it as a sugar-coated fairy tale. Academies of science that were not ornaments to ambition, but indispensable institutes for scholars concerned with the destiny of humanity, were not mesmerized into the ranks of Darwinian evolution.

Recurrent impacts of Darwinism persist in discussions among

university professors. A letter written to a newspaper carried on a dispute that suggested Darwinian science tends to support political conservatism. The writer criticizes an opponent who asserted "that science can at best illuminate the anthropology of morals but cannot decide the 'morality of morals.' But the quotations . . . at most suggest that insights from evolutionary biology can help us see why certain social institutions (and laws) are needed to control human impulses that were once useful to the species but can now be a source of mischief" (Lund 1998).

In keeping with this theme a psychologist entered the written debate with another statement:

> *SJG's thesis is that science and Darwinism do not shed light on the morality of morals. [This statement resulted from an accusation that some arguments are made which allegedly confuse evolutionary adapting with moral worth reported by Nelson Lund.] But SJG contradicts himself by suggesting that evolutionary biology can supply insight to the anthropology of morals. If so, we will have a biology of morality. It will spell out the particular reproductive advantages to particular moral codes (Paul 1998).*

> *For example, monogamy promotes cohesion by minimizing conflict related to competition for mates and sexual jealousy. Mr. Gould asserts that it is for religion rather than science to search out moral order. But if moralizing conveys a selective advantage on those promulgating the moral code, religions are likely to proclaim a moral order that conveys advantage to at least some of its members (Paul 1998).*

From professorial lecterns come words indicating that Darwinism has prompted bioscientists to compose their own forms of morality and ethical codes. Religion is brought into such discussions often to the diminishment of ecclesial dignity, respect, and honor that it had enjoyed prior to the Darwinian origin of species. Theories of human evolution account for heated scientific and religious debate recorded especially from 1844 to 1944 (Bowler 1986). The moral integer has eclipsed Darwinism to some degree by the anti-Darwinians in the decades around 1900. In a well-written book an author establishes the context and the necessity for reconciliation between Darwinism and the other biological subdisciplines whose alienation occurred in the years following the 1920s (Bowler 1992).

Confrontation with morality in Darwinism was aroused by Stephen Jay Gould, a professor of paleontology at Harvard, whose book, *Questioning the Millennium,* has given him notoriety in the academic world. He writes that he has been amused and also slightly chagrined by some conservative intellectuals. Accusing them of "involving the primary icon of my professional world—Charles Darwin—as either a scourge or an ally in support of cherished doctrines" (Gould 1998), he has stirred an uproar in academia.

He writes: "Since Darwin cannot logically fulfill both roles at the same time, and since the fact of evolution in general (and the theory of natural selection in particular) cannot legitimately buttress any particular moral or social philosophy in any case, I'm confident that this greatest of all biologists will remain silent no matter how loudly conservatives may summon him" (Gould 1998).

Quoting more of the paleontologist's views is an enlightenment. "The scourging of Darwin—the idea that if we can drive him away, then we can awaken, has animated a religious faction that views an old-style Christian revival as central to a stable and well-ordered polity" (Gould 1998). This quoted tirade is a response to writings by former Judge Robert Bork who stated: "The major obstacle to a religious renewal is the intellectual classes, who believe that science has left atheism as the only respectable intellectual stance. Freud, Marx, and Darwin, according to the conventional account, routed the believers. Freud and Marx are no longer taken as irrefutable by intellectuals, and now it appears to be Darwin's turn to undergo devaluation" (Bork quoted by Gould 1998).

There seems to be a contingency challenge that states, "If Mr. Bork will give me a glimpse of that famous pillar of salt on the outskirts of Gomorrah [Biblical reference to Lot's wife who turned into salt], I shall be happy, in return, to show him the abundant evidence we possess of intermediary fossils in major evolutionary transitions" (Gould 1998). The less vivid condition seems illogical, therefore it is preferable to consider it as an anecdotal metaphor. Controversy is inflamed by statements such as: "Perhaps I should be flattered that my own field of evolutionary biology has usurped the position held by Cosmology in former centuries, and by Freudianism earlier in our own times, as the science deemed most immediately relevant to deep questions about the meaning of our lives" (Gould 1998). And "embracing Darwin, many early-twentieth-century liberals lauded reproduction among the gifted, while discouraging procreation among the supposedly unfit. But science can never decide the morality of morals" (Gould 1998). When science advocates selective reproduction, is that not a moral decision on the morality of morals? The basic laws of morality seem to be discarded.

The publication by Gould may be a great literary event to the author but it contains a paucity of scientific and philosophic validity. Opinions stated are not warranted facts. Criteria must be extant before such statements as are made: "The abundant evidence we possess of intermediary fossils in major evolutionary transitions—mammals from reptiles, whales from terrestrial forebears, humans from apelike ancestors" (Gould 1998). The ape-human origin contradicts the Bible overtly. It is written, "And the Lord God formed man of the dust of the ground, and breathed into his nostrils the breath of life, and man became a living soul" (Genesis 2:7). Not acceptable is the belief of some scientists that the human living principle is called the soul.

The archetype fault in evolution lies not in the theory but its lack of credible, proven, documented, intellectual support. Without scientific glow it is diminished to a dismal unproven theory. Probably the majority of bioscientists think of evolution as either a matter of words and catchy phrases or as something too abstruse for the average mind to grasp and comprehend. Many aspects of evolution are like dry bones—a skeleton—without muscles, integument, and *sans* vitality. Bereft of essential scientific elements, evolution does not emanate the inspirational sentiment attributed to other scientific proposals of the same era.

Chapter 17

Contradictions

Revealing studies on the lifestyles of native wood frogs and turtle species in Canada and the northern United States indicate they have an uncanny ability to withstand cold temperatures and freezing weather. Investigators found to their amazement that wood frogs go without eating for an entire winter. Moreover, "They spend those months frozen solid—without a breath, without a heartbeat, without measurable brain activity. One day they are no more than frogsicles; the next day [defrosted] they are too nimble and squirmy to catch" (Storey 1999).

The enigma intensifies when thought is given to bears who hibernate on land, whereas common box turtles, leopard frogs, or bullfrogs retreat underwater during the winter. Wood frogs were studied more easily by experts in amphibian zoology who evaluated spring peepers, chorus frogs, and gray tree frogs as well. All survived a winter freeze, and super-cooled surroundings. Even when frozen stiff like icicles, appearing to be dead, some animals can survive the harshest winters. This is a magnificent, almost puzzling observation when it is realized that botanists warn that even a mild frost can kill hardy plants. Zoologists teach that animals adopt one of two strategies to deal with severe cold. They either tolerate freezing or they avoid it at all costs (Storey 1999).

When victimized by cold temperatures the liquid in animal cells "crystallizes and expands as it freezes, breaking up delicate tissues bursting the walls of blood vessels" (Storey 1999). This causes cells to crenate with an accompanying disorganization of their intricate cytoarchitecture. Human nature is such that people are accustomed to warm bodies. Thus they associate warmth with life and cold with death. Only birds, certain mammals, frogs, fish, and perhaps a few other categories of undiscovered species can generate sufficient metabolic heat to survive intense cold by elevating their bodies to a higher temperature. *Homo sapiens* has not this luxury. Evidently this physiological capacity argues for an identifiable biological asset not found in human beings, which in this regard categories the specifically mentioned animals, frogs, birds, and others as superior to them.

This bioscientific fact is not a boost to the theory of evolution. Is this a reversal in the speculation of the survival of the fittest? The superiority of lower life over human beings is manifested also by cold-water marine fishes who have an internal antifreeze. "To keep from freezing in the Arctic winter some members of the smelt and greenling families synthesize enough glycerol to push their freezing points below twenty-eight degrees . . . Most fishes, however, use antifreeze proteins, not glycerol, to protect themselves against the cold . . . Antifreeze proteins work by binding directly to the growth plane of an ice crystal, breaking it up and inhibiting more water molecules from joining the crystal lattice" (Storey 1999).

Remarkable research such as these studies is illustrative of the frailty of humanity by comparison. Humankind's intellect and free will are attributes of the soul indigenous to it. The existence of the soul is not accepted universally by ardent Darwinians who have denied creation. With this superiority is an inferiority. One proven deficiency is the inability of human beings to have access to an internal antifreeze that can prevent deadly hypothermia. Unclothed human beings have an impediment to the survival of the fittest theme if exposed to temperatures below freezing.

Greatness is a matter of consensus. But "Fame is like a river, that beareth up things light and swollen, and drowns things weighty and solid" (Francis Bacon, 1561–1626). And Darwinism was not and is not now an encouragement to moral ascendancy. When an opined theory collides with conventional wisdom, not necessarily religious, a controversy of incredible proportions is anticipated.

Darwinians react to negative, destructive criticism as merely the end of a chapter in the total text on their believed-in evolution. Failure to recognize the impermanence of the theory is an indication of an intellectual inability to differentiate phenotype from genotype. What is visible: the environmentally and genetically determined observable appearance of an organism is phenotype. The genetic constitution of an organism especially as distinguished from its physical appearance is genotype, meaning what an entity really is.

In their fanatical advancements of a theoretical, ideological agenda, evolutionists blithely belittle critics as biological infidels. Acrid oral and written attacks are made against cogitative academicians, fair-minded scholars, and eminent anthropologists who are in disagreement with survival of the fittest.

Darwinism appears to be the process for the stabilization of an inter-connectedness among the components of the animal-human species that amalgamates them into a continuous heredity. However the congenial appearance of the theory, the conviction of its postulants, and the zeal of its adherents give neither correct, scientific strength nor grant it absolute truth.

Although valued by a consortium of bioscientists every expert, disciplinary, scientific field of learning does not embrace the speculation wholeheartedly. Equivalent, appropriate, erudite persons of professional caliber in social, cultural, humanistic, anthropological, and philosophical circles have doubt about evolution's authenticity. The academic scientific society of scholars is not a monologue in support of the *Origin of Species.*

Darwinian evolution is under reconsideration by a coterie of distinctive observers. Although the theory does not seem to be infinitely complex, its systemic interdependence is not elucidated with clarity, facile comprehension, and satisfactory substantiation. The paradigmatic revolution that erupted out of the Darwinian theory is identified as the product of either an observed or deduced anomie that deviated from official, coded, biological criteria. Unsustained interpretations, models, and similar adjuvants to the theory must be modified and find more validity to be acceptable. Otherwise it is to be abandoned as a biological validity. Scientific viable communication requires directional accuracy. Exercising intellectual power of persuasion without firm, basic, authoritative proof devalues the project advocated. In spite of apparent imperfections Darwinism has its adherents who are magnetized by a concocted tale that breeds celebrity status to those who expound it.

Prudent flackmeisters, considered harmless, created the illusion of species continuity by survival of the strongest. Prior to exploitation of the public by exhibitors of curiosities, the world had not seen science as entertainment. Perishable, fictitious, hyped-up advertisements increased notoriety which intensified interest in evolution as an attractive business commodity. Celebrity ceased to mean famous as it converted fame into business value, money, and persuasive star power. Formulated into entertainment, evolution did not require the need for those prestigious elements that verify scientific certainty with its firm imposition of austerity against frivolous falsity.

Chapter 18

Creation Theology versus Theory of Evolution

Many objectors to evolution remain silent, fearing retaliation by persons with higher academic standing, administrative authority, and/or punitive power to expel them from prestigious professional societies. Bioscientists of great stature have not erred on the side of error. They have compromised their keen, intellectual integrity by aiding in the insinuation of evolution doctrines into science's dogmatic culture. Their contentious arguments could be the kindling fuel to ignite a conflagration that could destroy the sustainability of honest science.

Any extreme change in bioscience requires great courage, excessive discipline, must not bring with it chaos, and not stimulate resentment. Academic liberators known for their eagerness to uplift freedom of the human spirit from all restrictions inadvertently suffuse hatred whenever there is dampening of the freedom desired.

Flights of thought in evolutionists were not repressed as they became entwined with the ongoing cultural formation of the last decades in the nineteenth century. To be *au courant,* evolution was absorbed easily by non-scientists whose overly generous enthusiasm was a gigantic encouragement to Darwinists. No one gave any attention to the intrinsic value of the theory as having scientific truth.

An exaggerated theory has no valuable research documentation that sets forth without proof fundamental principles that are accepted without disputation. Validation is made memorable of that which is an accurate and incontrovertible claim. Veracity always is uplifting even for the downfallen. Well-documented bioscientific research with provable, reasoned conclusions is an impressive inspiration to other researchers who with foresight and insight seek similar discoveries. By mental inspiration the spirit is energized to ensure merited recognition while enhancing the probability for a high award of scientific recognition and world prestige.

Sad is the commentary when inaccuracy is elevated to celebrity status by virtue of one-sided, prejudicial publicity while more valuable ventures are unsung. Speculation is irresistible to an active mind. Without some resonance of magisterial feasible approbation in a hypothesis there is minimal hope for acceptance, whereas attestation can turn around an initial misbelief into a probability that encourages additional studies. Thus pivotal contributions can be made that may find answers for what is lacking. Scientific proposals encumbered by doubt arouse cynicism with a failure of acceptance anticipated.

Sometimes touching on the fantasy, evolution obviously has neither a contact nor a contract with God. No misconception is the indication that to some acute observers, evolution idolizes atheism. Anthropologists, bioscientists, and sociologists know that evolutionists are brainy and feisty. None of them are contradictory to one another and self-effacing. With all their combined brilliances they have failed to supply the alien hybrids to fill in the empty spaces in the necessary serial connecting links and biological concatenation that would mandate acceptance of Darwinism.

Weakness in evolution is found in the unsatisfying absence of answers to questions in the void of a harmonious continuation of selectivity for the survival of the fittest. As of this day there is no vital information provided on this specific, undiscussed linkage topic. There is no documented text that completes it as a natural law of evolution. Obvious obscurities add to the lack of explanation that increase the darkness surrounding the Darwinian theory.

The golden years of evolution have been tarnished by the disease of incompleteness. Although the illness is not gone, the theory is far from the crematorium. Having had a temperate remission after a prolonged recrudescence in the twentieth century, the twenty-first century augurs a revival among the more recent ranting guardians of its tenets in academia. Without gaining a modicum of vigilant scientific validity, Darwinism is discussed favorably on university campuses. Many bioscientists have their weapons aimed at divine creation. During this current period of reawakening, opinions are expressed that do not add one *iota* to what exists.

Controversial evolution is recognized as an innovation. But it is neither focused on nor is it equal to epitomized bioscience. Without defined accuracy, Darwinism does not include unlimited, reasoned certitude that clarifies all ambiguities. Demise of logic, breakdown of rational continuity, miscalculation, and no triumphant progress deny evolution enshrinement as a seminal contribution to the biosciences.

Evolution had the trendy design for the inquisitive minds of the 1860s that illuminated musing imaginations during their nonproductive hours. Also prompted was the assertions for defense of Darwin who was at his home like a recluse enjoying the leisure of satisfaction. In the meantime, other known personalities were his human broadcasting system. Darwin obtained the seal of chic approval from gullible laity and adoring contemporary biologists. By their certification of Darwin his theory was given credence. In turn the commercial exhibitors and their collectibles were believed.

Simultaneously, the core of the subject was not scrutinized as to truth among consumers of the survival of the fittest proposition. Strange to relate, those accepting Darwinism and its independent promoters did not question the details of the omitted aspects of evolution. Procedures used to arrive at conclusions presented are not explained. No mention is made of the use of traditional scientific methods. Not revealed is that pressure from paid promoters was the propulsion that fueled the madness for the unproven doctrine. Each promoter had his own claim to fame. Hence the social stature and reputation of them prompted agreement with evolution. Public acceptance surpassed all anticipation.

Being well trained in experienced, elegant style with elite venues enabled some promoters to meet business dignitaries. Brought about thereby was promised entree into the most influential and distinguished social circles. Offered opportunities became available to publicize Darwinism in large and small cities. This was a factor in the bothersome issue weathering the storms of criticism. Hence the theory remained popular even when the clergy tried to correct, even punish, the proselytized evolution's agnosticism. Hostility became a fledgling endeavor in comparison to the enthusiastic response to the theory.

As a very articulate group, evolutionists were qualified to function as evangelists for Darwinism. No imposed isolation, no great struggle, and no lonely heartache was theirs to suffer. Addressing the multiple challenges adroitly the dramatic proponents solidified rather than destroyed the theory without any financial disappointment and/or loss to the exhibitors. Opponents to evolution, not its promoters, remained the empty reminders of the vigorous, able leaders they had been formerly. Darwinists encouraged each other to be assertive. Often they found themselves in places of academic leadership and educational responsibility that they accepted, obtained, and/or were invited to assume.

These circumstances produced an environment that provided additional opportunities to express approval of the origin of species. With wide recognition, the tenets of evolution were advanced. Academia gave Darwinists fortunate positions that stylized their presentations authoritatively. Images created were dramatic, concentrated, and educationally hypertrophied. The pleasant period for evolutionists continued for a long time. They made bioscientific statements which motivated many people of almost every genre to visit commercial exhibitions, participate in scholarly studies, and listen to erudite speakers. Ready-made audiences increased, which added more attractiveness to evolution's charisma. Curiosity cabinet exhibitors brought their shows to urban areas far away. Rewarding experiences in familiar atmospheres encouraged showmen to seek international exposure for their memorabilia. All the propaganda collectively gave evolution a glittering that made it the most talked about problematic subject among middle-class society and those in the social register.

Darwinian colleagues did not live in a communal compound. Neither were they all formal associates nor were they united in social relationships. Each was his own agent accepting lucrative lecture tours that produced a substantial income which was more than temporary. Their opposing scientific contemporaries could not match their financial worth, even though they may have contributed much more real knowledge to the scientific culture of the same era. This proves the financial success of elaborate advertising as an effective way to draw attention. The array of evolutionists did this very well. Added to the expertise was the inspired shopkeepers of the collected specimens whose classy, classical styles and craftsmanship kept alive lavishly with decorative motifs the fashionable interest. These displays were the hallmark of evolution becoming unique to it.

The history of evolution provides a review of it. Minute evaluation suggests it is not a written record of a growing body of bioscientific knowledge but rather a discourse on the strenuous efforts made by Darwinists to spread their convictions. The dogma of evolution offers a beginning induction without a proven, verifiable ending to a long, exciting proposal. Lacking a featured treasure it is not accepted as a *non-pareil* contribution to anthropology, biology, and sociology.

The splendor of the nineteenth-century supporters of Darwin was demonstrated by the refinements that supported it. This included those highly respected in literature, science, upper-class society, and royalty. Meanwhile the numerous curiosity collections became Mecca for thousands of thrill seekers. Not only did they draw the largest attendance at each session, probably they were the best non-performing, slickest show available.

People responded very well and favorably to the objects offered for staring review during the mega-events. Like a mini-circus, the material viewed was exhilarating, mesmerizing, and fascinating. Nothing presented was scientifically informative nor had proof-positive to support the truth of evolution. Seeing an assortment of old things in new surroundings became a novelty of pristine oddities. Mingling crowds of the unsophisticated were paying handsomely for the privilege, according to unscrupulous vendors, to look upon what was often a puzzle to them.

The Kansas State Board of Education voted 6 to 4 in mid-August 1999 "not to include a central strand of Darwinism theory, so-called 'macroevolution,' among its suggested standards for statewide high school testing." American education was irked into an uproar. Joining in the commotion were many civic groups and eager legal beavers.

Creationism is not mentioned in the decision; it does not promote religion, and does not advance the biblical story of Adam and Eve. "The board's decision changes no state educational policy; local districts will continue as before, to decide on their own curricula—including Darwinism." The vote has no impact on the U.S. Supreme Court ruling that prohibits schools from teaching creationism as science.

"What the school board has done, in attempting to define a body of scientific knowledge with which Kansans should leave high school, is to omit from its suggested list of topics the Darwinian theory of macroevolution—which has it that one species evolved from another." The board has retained, it should be said, the equally Darwinian theories of "microevolution" to explain changes occurring within the same species, and "natural selection," the idea that improvements to a species occur over time, as well-equipped creatures survive and propagate.

For all its pleasingly cogent arguments, Darwinism fails to answer intriguing questions, among which is how major animal groups appear abruptly in the fossil record, rather than gradually taking shape over the ages as Charles Darwin would have predicted. Phillip E. Johnson, a law professor at the University of California at Berkeley, related an anecdote in the *Wall Street Journal* about a Chinese paleontologist whose lectures about fossilized inconsistencies with Darwin's theories affronted American scientists. The Chinese scientist was moved to observe, "In China we can criticize Darwin but not the government. In America you can criticize the government but not Darwin" (Editorial, *Washington Times* 1999).

Emergence of newer concepts of design in the sciences marks the return of final causality to bioscience. A broad relevance to the notion of "black box" as a purposive open system whose inputs and outputs are observable and systematically related has been recommended (Behe 1996). *Black box* is defined as a device or theoretical construct with known or specified performance characteristics but with unknown or unspecified constituents and means of operation *(American Heritage Dictionary,* Third Edition). Challenges to Darwinism are recurrent. Confusion in science will be eliminated as the new intellectual forces strive for global homogenization in terms of free interchange of ideas, knowledge, and bioscientific conclusions.

Not surprisingly is that the allies of evolutionists will denigrate any person and evidence, no matter how compelling, that challenges the theoretical constituents of evolution. Beneath the acrimony is the central atheistic icon of a stringent, dogmatic denial that is intolerant of any divine participation in the biosciences. Darwinians illiteracy of God allows their own indiscretion to be their tutor.

When the French scientist Lamarck published his thesis in

1809, it was written to explain what he believed to be the mechanism of evolution. This was two years prior to the birth of Charles Darwin. Lamarck had few colleagues who accepted his explanation and a lesser number of scientists who followed his speculation. Many years later he had more exponents who believed that species development *via* use and disuse of anatomic structures was part of the survival hypothesis.

The underlying theory of Lamarck and Darwin was their observation of what they believed to be factual. Namely that existing species developed from other preexisting species with their added opinion as to its occurrence. Foes of evolution point out flaws in the theory as to the development of species. Their conclusion is a negation of it (Futavama 1942).

However, from the research on developmental embryology, a relationship of certain structures in different species have a similarity which cannot be ignored. Biologic homology in different species is so close as to require a valid explanation (Filica 1995).

The eminent convert, John Henry Cardinal Newman, in 1862, when the evolution controversy was fulminating in Protestant England, remarked that in view of the morphological similarity between men and apes, the *onus probandi** rested on those who denied, rather than those who affirmed the existence of a genetic connection between them (Johnson 1948). Although the human mind seeks to find the cause of these observations, explaining how one species of life could evolve from a primitive, rather simpler form into a complex, physiologic mechanism is most difficult.

Too many scientists become so deeply involved in their career projects that they are not concerned with outside activities. One scientist of professorial recognition stated that he only reads scientific journals on genetics which is his specialty. Since his student days he has not read the Bible, a newspaper, literature, or magazine. Added to this remark was the statement he "had no time for religion." This professor is not alone in this sort of *via vitae*. The number of his professional colleagues of similar enterprise is not countable. If major scientists are not concerned with that which is visible, discernible, and legible, how can they be persuaded to turn to God, who is invisible?

Although scientists have in their hearts a great ideal that drives them to pursue authentic values, which should give solid meaning to their lives, they are missing the source of antidotal consolation when failure confronts them. Unpleasant disappointments and insufficient concrete achievements are unburdened when God is a scientist's companion. A scientific journey either with or without success, made in God's company, is a horizon of joy. It gives deep, rich gladness, and hope for future high attainments with exhilarated confidence.

Social scientists for a long time have intimated that religion is an antique vestige that is trying to survive in an upgraded world of elegant, sophisticated technology. With positive anticipation they await the date of religion's demise. They postulate that as science advances so spectacularly, religion diminishes unwept, unhonored, and unsung. Furthermore, sociologists teach that the more educated people become, the less religious they are. Such assertions are not correct.

"Despite the explosive growth of science and the increase in average education levels during the last half-century, the rates of religious belief and participation in the United States have stayed about the same. It is true . . . after examining extensive surveys from the period 1972–90, that professors and scientists are less religious than the general public . . . Outright rejection of religion remains more common among academics, however, but that may be because the irreligious are more drawn to the academic life, not because higher education reduces religious belief" (Iannaccone, Stark, Finke 1999).

Not all sociobiologists seem to support evolution as the answer for that minor segment of the intelligentsia world which denies creation categorically. This conscious choice is an asset for the proponents of materialistic relativism. Unprincipled pragmatism, collated with evolutionary determinism, aims at the destruction of theological creationism down to plausible nihilism.

Scientific minds closed to God appear to have abandoned logic and have embraced militant atheism in united attacks on natural law. Forcibly this procrustean didacticism has grasped agnostic relativism in defiance of divine truth. Evolutionists have confined their profession to a monologue not to a protean career. With much dexterity they overtly display a sense of superiority and confidence in their own rectitude.

Emblematic of underwhelming evolution is its abject disavowal of human creation. Like a fasces minus a ferrule it has no reinforcement. Without an omniscient, infinite divinity as sustaining sustenance, evolution is doomed to finitude. No countermand, criticism, denunciation, diatribe, and invective can either alter or subdue this conclusive, severe reproof with strong censure. A dying theory can never fend off the warmth of creation that removes the dawn chill of falsity.

The best periods in history have been those when Faith and Reason were of equal weight and when they saw each other from a respectful distance. Faith kept retreating, finally exiting the drawing room of polite discussion in the nineteenth century. It was pushed aside by Darwin, then buried by Huxley and Nietzsche. At the cellular level, publications by Lehigh University's Michael Behe and others point to design in evolution, as opposed to the long-held theory that evolution is the result of arbitrary mutations. Design implies a designer: God.

Chapter 19

Darwin and the Descent of Mankind

In chapter 9 of a remarkable book with the dramatic title, *A World without Heroes: The Modern Tragedy*, evolution is labeled a fictional uncertainty (Roche 1987). The author writes, "The distinguished philosopher Mortimer Adler once termed evolution a popular myth, which is just what it is . . . The evolutionary epic is the best myth we shall ever have affirms sociobiologist E. O. Wilson, professor of science at Harvard University. Evolution has been called a curiosity that reflects human gullibility" (Roche 1987).

Certainly it must be admitted this subject has exuded more excited pseudo-scientific banter than any other topic during the past century in the English-speaking world. The intellectual exchange is kept burning as it is refueled mainly by theoretical scientists. Hence the furor over evolutionist doctrines has not diminished since the incipiency of Darwinism.

"To the detached observer, the ongoing feud between creationists and evolutionists has all the marks of an unholy war between two fundamental sects. Even the scientific aura surrounding the question is invoked for clearly religious reasons. To the anti-heroic faithful, science itself is quasi-divine and its revelatory gospel is not to be questioned" (Roche 1987). Religion and science are not and should not be considered to be paradoxically contradictory and hostile.

Evolutionists implied by their proposals that the swathed scientific dogmas of the slathered, tranquil past were anthropologically inadequate. The forceful attempts to advance evolution as factual disenthralled many bioscientists. Those who had a penchant for medical history foresaw that future historians would record Darwin as an atheistic enemy within the ranks of scientists.

Most dedicated scientists rely upon their native intelligence, innate wisdom, and acquired education to achieve their goals. These are blemished when they do not have an upright moral character and a monitoring conscience. Hence there is a religious segment in the deeds of every learned person pursuing science as a discipline. Good behavior in their endeavors includes honesty, integrity, probity in the laboratory, no deception in observation, admission of failure, and modesty in success. A motto of value is "To thine own self be true and you canst not be false to any man" (Polonius to his son, Laertes, in Shakespeare's *Hamlet).*

"Evolution is and always has been a religious vision. It is a vision that denies God, but is not less mystical or mythic on that account. Evolutionism must be taken on faith and that is its major support. The only relevant question about it, in the end, is whether it is the true faith" (Roche 1987). Evolution is no more than a dubious hypothesis for scientists to defend. No simple test, experiment, archeological discovery, or anthropological study sustains it as a science. If any upholding evidence existed in past ages it has left no identifiable trace. These missing elements immune evolution against noxious questioning by scientific-testing. Until such procedural scrutiny is possible, nothing of value will be available to champion it as something more than a theory. Rigorous, sturdy potency for evolution has not been forthcoming from the cladists, the scientists who classify fossils. "A minority of respected scientists think evolution is flatly impossible on other scientific grounds" (Roche 1987).

Darwinists admit that life has many unsolved mysteries. But they insist evolution is a fact. Factualization of evolution was cited by the defense during the highly publicized Scopes monkey trial in 1925 that brought fame to Tennessee. With all the rhetoric that filled the courtroom no positive proof was offered, no scientific evidence was presented to bolster the claim that evolution could be objectively verified. No reference was made to evolution as a deduction from the theory of natural selection. It was not maintained by any of the legalists and none of the expert witnesses that evolution was a discovery from demonstrative evidence and thus remained unproven.

Gilbert Keith Chesterton (1874–1936), the English man of letters, expressed his thoughts on Darwinism when he wrote: "Evolution is a good example of that modern intelligence which, if it destroys anything, destroys itself. Evolution is either an innocent scientific description of how certain earthly things come about; or, if it is anything more than this, it is an attack upon thought itself" (Roche 1987). Of a certainty the master-minded Chestertonian writing on evolution was not a colportage.

Nature moves toward perfection, said Darwin. Humanity draws inexorably nigh to a socialist heaven on earth, said Karl Marx, who lived from 1818 to 1883. When the German philosopher, political economist, and founder of world communism asked to dedicate *Das Kapital* to Darwin, the evolutionist refused (Roche 1987). Nature and humanity by Darwinian and Marxist standards were the popular ascendant mental images in the mid-nineteenth century. Acceptance of Darwinism and Marxism became potent forces toward the contribution of thoughts for social change. Both were powerful weapons against the wounded, waning Christianity which was being routed systematically out of the civilized world. Without steadfast adherence to Christian precepts, ethical conduct and moral behavior declined.

Readers of the *New York Review of Books* learned that Steven Jay Gould is angry. That was in June 1997, when they read a tirade in which Gould tried to settle a disagreement with some of his more prominent guild of Darwinists. The war was long standing, erupting periodically for almost two decades.

In the early 1980s, the British geneticist J. R. G. Turner

remarked with specific reference to the controversial Gould that evolutionary biologists are all Darwinists, as all Christians follow Christ and all communists Karl Marx. Turner declared the schisms to be over, which allowed parts of the master's teachings to be seen as central. "Gould, like Gorbachev, deserves immense credit for bringing glasnost to a closed society of dogmatists. Like Gorbachev, Gould lives on as a sad reminder of what happened to those who lack the nerve to make a clean break with a dying theory" (Johnson 1998).

Scientists have the ultimate goal in their research projects to seek provable truth and preserve it. The minds and hearts of bioscientists, their co-workers, and colleagues by training are dedicated to finding truth in its totality: scientific, theological, and practical. Human achievements and accelerated progress in scientific and technical fields must not fall short of moral subjectivism. Honesty in science plus diligence against falsity disallow deviation from those ethical values that cause improper views of reality. Disruptive immoral forces lead to the inescapable ineffable poverty of scientific advancement. Evil intentions imperil the stewardship of altruistic science because such deliberate purposes transform it into an insidious, unholy commitment to evil. Misuse of intelligence is a misalliance with humanitarianism.

A scientist, like any other moral person must be faithful to the unwritten law of conscience. Beguilement is not connate to science. Integrity is the basis for dignified human public deportment, personal private behavior, and societal conduct. Adherence to a rightly educated conscience does not tolerate exploitation of a person for any feigned excuse or conjured reason. Reclamation by scientists of bioethical values will forge an unbreakable bond between science and honor that will make them guardians of truth.

Many biomedical scientists are looked upon by their colleagues as extremists when disagreement occurs. Sometimes each side can become unreasonable as they may be blinded by an ideology that makes them obtuse, recalcitrant, and unyielding.

Eventually exaggeration increases the magnitude of the differences between them which creates bitter adversaries. However, dissent does not hurt careers, although personality conflicts may flair into animosity and envy.

Psychology tries to explain this diminution in expected conventional wisdom. In-depth studies reveal quite often that the demonstrable antipathy is not as extreme as opponents estimate. Psychologists use the term *naive realism,* defined as the tendency of people to believe they see the world objectively and "anybody who sees things differently than we do must have judged that issue or object through some ideological lens" (Keltner and Robinson 1997). It is most applicable to both sides of Darwinism.

Another term used to explain excessiveness in opinion is *lone moderate* (Morin 1998). This centers on those persons who think that even those who share the same belief are more extreme than the selected individual rendering the same conclusion. This phenomenon permits a debatable deduction. It is that "[p]eople sense they alone have got the facts right, that they are the balance between extreme opinions" (Keltner 1997).

Although judgment may decide one is a moderate, partisan rivalry persists. Opponents remain "hostile, irrational, immoral, and ideologically extreme. If most people assume that ideological debates are matters of black and white as extremists stand in opposition to each other, then moderates and moderate solutions may not be heard" (Keltner 1997).

It is thus that spokespeople chosen on both sides of an issue are extremists because they assumed that position was typical of the majority when really it was not (Keltner 1997). This may be a self-fulfilling prophecy of strong advocates for a cause that glorifies the person more than the advocacy. Darwinian evolution fits into this discerned pattern.

Chapter 20

Evolution Science and Creation Theology

From the repetitive dialogue of the thorny discourses, debates, and disagreements on religion and science emerges periodically the volcanic fumes of Darwinism. Unpredicted is the subclinical rebellion against religion that arises occasionally in the usually tranquil, settled academic ambiance. Scientific faculties at universities reconceptualized the teachings of their predecessors in search of newer truths. Religion no longer is the fortress against educational uncertainties. This ushers in a deconstruction which can be defined variously as disruption, transgression, undermining, a dangerous restudy, and a risky undertaking (Caputo 1997). Some faculty advisers of doctoral candidates in European universities claim that what was absent in a thesis had more authority than what was written.

Denunciators of socioeconomic standards by certain groups of men and women have incited intolerance to academic tradition. In anthropology and other bioscientific disciplines it is fashionable to attract listeners to trash the concept that human beings share a common biologic nature. Often the belief in human creation is dismissed with irate contempt by tenured professors.

Some have coauthored that: "To set humans apart from even our closest animal relatives as the one species that is exempt from the influence of biology is to suggest that we do indeed possess a defining essence, and that it is defined by our unique and miraculous freedom from biology" (Ehrenreich and McIntosh 1997).

This outlook, they observe, is "eerily similar" to that of the fundamentalist creationists now waging war on the theory of evolution (*Wilson Quarterly* 1997).

The existing conflict between science and religionn has intensified. Reawakened was the reposed trial of Galileo. This key episode of the seventeenth century was used to discredit biblical teachings (Feldhay 1997). The Galilean inquisition heightened the verbiage on the science-religion conflict until Darwinism, which became more quantitatively persuasive from the nineteenth century onward.

Many scientists have had various degrees of power which influenced intellectual and societal thought as a potent force to challenge religious tenets and authority. The British naturalist, Charles R. Darwin, with his theory of evolution promulgated the concept of natural selection by survival of the fittest. By his origin of species he instituted a battle between those who subscribe to his theory and those who believe in humanity's divine creation. Darwin's publication dealt such a blow to the Bible that "it made it impossible for any conscientious and thoughtful man to accept—as he had still been able in 1800—the Bible as literally true" (Roberts 1993).

Darwinism emboldened shopkeepers of human spermatozoa, those bioengineers of cross-race artificial insemination, baboon xenograft enthusiasts, and guards of ova imprisoned in petri dishes to be proclaimed as bioscientific heroes. It is postulated that the belief in Darwin's origin of species encouraged twentieth-century scientists, released from all religious restrictions and bioethical constraints, to manipulate human genes.

The underlying impetus to genetic engineering may find its origin in Darwinism. This is open to question. Some do believe that the purpose and goal of the Human Genome Project is to sustain the theory of evolution with secondary benefits of genetic screening and counseling.* Raised inquiries for which answers are sought range from the essence of what it means to be human to the validity of Charles Darwin's theory of natural selection of the fittest organisms through their greater reproductive success. The issue of genetic quality has been the topic of discussion, debate, and disagreement among many authorities in bioscience, theology, philosophy, sociology, and law for many decades. Reproductive genetic engineering has evolved as a current contentious issue.

Acquiring an eminent reputation by his promulgation of the shocking theory, Darwin aroused bioscientists in a most peculiar manner. But the high commendation of the author was not promoted by irrefutable proof and without fidelity to scientific verification. Rejected by many of the learned as spurious, accusations of extravagant perfidiousness were heard. Retaliation alleged calumny by the envious. No malice was intended by all participants.

Certainly no sanctioning sacerdotal formal discourses occurred. Contrarily, Darwinism was considered intellectually ignoble by many religionists of diverse denominations. In all genre of people there were those who found the theory one of disbelief in the divine. The deducted perception was one of dismissing God from the origin of humanity. As a socio-historical innovation, evolution was an attempt to establish a biological genealogy.

Human beings since the beginning of time have practiced some form of religion. Individually and in groups, people of every description have lived without adequate food, money, clothing, shelter, education, transportation, laws, and other things now deemed necessaries, but not without religion. As science grew into a discipline of higher learning, religion diminished in prestige. Insidiously, it was transmogrified into divergent and often dissimilar creeds. Man-made interpretations of religion produced a decline in its overall authority to invoke "Thou shalt not." Because of its many divisions and subdivisions it lost its influence, persuasiveness, respect, and command accession. One part is never stronger than the whole.

Romance with God seemed to be over. In this climate creationism was doubted. To some it was a foolish belief, as loss of religious faith became more visible. Making fun of religion was not a rarity. "H. L. Mencken* and other thinkers once scorned religion as akin to imbecility. Today's intellectuals have abandoned the tradition of caustic secularism that once provided refuge for the faithless" (Kaminer 1996). A resurgence in non-iconoclasm is on the rise as the twenty-first century is awakening with beneficial expectations.

Several years ago the president of Queen's College University of Cambridge, a particle physicist and an ordained Anglican priest, wrote about a very curious coincidence. "When we use mathematics as a key to unlock the secrets of the universe, something very peculiar is happening. Mathematics is the free exploration of the human mind . . . Inexplicably, some of the most beautiful patterns thought up by the mathematicians are found actually to occur in the structure of the physical world. In other words, there is some deep-seated relationship between the reason within (the rationality of our minds—in this case mathematics) and the reason without (the rational order and structure of the physical world around us). The two fit together like a pair of gloves" (Polkinghorne 1996). Religion says that the reason within and the reason without have a common origin in a deeper rationality which is the Creator. Theology has the power to answer a question, namely the intelligibility of the world, that arises from science but goes beyond science's ability to answer. Science assumes the intelligibility of the world. "Theology can take that striking fact and make it profoundly comprehensible" (Polkinghorne 1996).

Authentic Christianity and the world, including scientists, have been at odds. To regain popularity, to entice parishioners back into the congregation, and render religion easier to take, changes have been made by religious administrations. Religion has become easy, upbeat, convenient, palatable and compatible with secularism. It does not require self-sacrifice, discipline, humility, an unworldly outlook, a zeal for souls, a fear as well as love of God. "There is little guilt and no punishment, and the payoff in heaven is virtually certain" (Reeves 1996). Although all these opinions are not universally accepted, it is admitted that laxity in some Christian denominations does prevail. This is a debility that does not enforce faith against scientific error even as it lessens the gravity of sin.

Religion separates people from all other living species and is not a mere transient space in human progress. Contemporary civilization records that religion not only carries on but is flourishing in some countries. Evolutionary biology has made irreligious converts by its proponents who have rehearsed their banal theme with the desired effect. Serious thinking scholars doubt evolutionary science. They do not accept it *carte blanche.* In deeper thoughts they find fault when one tries "to postulate a definite inherited program of behavior, encoded genetically and passed on in continuous evolution from more primitive to higher living beings and culminating in man. The examples from different species are not connected by a continuous chain of evolution . . . We are dealing with analogies, not homologies" (Burkett 1996).

One of the best-savored ironies in the history of science is that Charles Darwin wrestled vainly for years with his theory's major problem while the solution was right in his own library. The problem was blending inheritance: why are not the features an organism inherits simply an equal mix of its parent's features, so that all the differences among individuals in a species eventually average out? "The solution—individual genes—was in Gregor Mendel's ground-breaking monograph on pea plants, a copy of which lay unappreciated in Darwin's study" (Dennett 1997). The evolutionist missed the clarification of his own implications by not opening a document written by a religious scientist which was within arm's reach. Perhaps the naturalist-botanist was prejudiced against a man of God. By ignoring Mendel's treatise, Darwin missed the cultural transmission of his would-be meme.* This adds to the pitfalls of simplistic Darwinism as it implies that "Darwin's breakthrough in biology grew out of his deep knowledge of a wealth of empirical details scrupulously garnered by hundreds of pre-Darwinian, non-Darwinian natural historians" (Dennett 1997). In these existing years of computer excellence, expertise, and endearment, it is highly improbable that artificial mechanical intelligence will unravel with satisfactory explanations the puzzling conundrum of the mystifying intricacy of evolution.

Persistent emphasis on the supposed conflict between creationism and evolution may be a scientist's delight. Serving no tangible, beneficial purpose, it makes for discord in a relationship that ignores the purpose of each which is to seek the ultimate criterion of truth. Certainly, without scientific discoveries and inventions, human, animal, and vegetative lives would not have improved. But science does not have all the answers to questions that arise. Every scientist knows or should know the limitations of his/her profession. Even the agnostic Sir Julian Huxley (1887–1975), British biologist and author, realized that "sensible man of science imagines for a moment that the scientific point of view is the only one. Art, literature, religion, and humane studies are other ways of exploring and describing the world." Catholic scientists defend the evangelization of creationism over Darwinism.

When the *Origin of Species* was published in 1859 the scientific world, by previous indoctrination by persuasive scientists such as Charles Lyell, was prepared for theories that were incompatible with Genesis. Darwin's publication enforced the anti-religious propensity into a positive activity. Literal reading of the Bible has continued to be attacked ever since by many scientists, reformers, and self-proclaimed zealots of freedom in thought, word, and action. By their silence some scientists gave tacit consent to evolution. Those who do not commit themselves to a cause and those who are neutral are an invisible majority. Perhaps they embrace the doctrine of political correctness for their own selfishness and survival. An indifferent intellectual attitude spoils the willingness of choice in decision-making, indicating a weakness of free will and lack of moral commitment.

Darwin, as well as many other botanists plus, zoologists and anthropologists, observed with recognition the multiple variations visible among living organisms. These are obvious in domesticated animals and plants seen in households. External features vary which augment beauty, giving indigenous attraction to each. Darwin did not know the reason for these changes even though Gregor Mendel had published his seminal experimental conclusions in 1865. This lack of knowledge was equated with his omission of deduction that variations in a phylum are never such as to form a new species. More acceptable is Darwin's discovery of how a species will adapt to different environments. A distinct flaw in evolution, totally ignored, is that deviations from the normal are in the direction of degeneration and do not represent an advancing, upgrading regeneration with augmentation (Felice 1996).

The weaknesses in Darwinism have been exposed vociferously by fundamentalist Christians. Their continual attacks will not cease until the theory of evolution is bludgeoned into the grave. The Christian opponents to this theory adhere tenaciously to the belief that natural selection is contrary to the basic law of the universe and to the Second Law of Thermodynamics. This latter law stipulates that every event in the universe leads to greater disorder and less usable energy (Felice 1996). Hence destruction is the process, not new construction. The large numbers of scientists betrothed to Darwinism have ignored obvious flaws in its theory because evolution is a powerful weapon against a common enemy: Christianity. Augmented vilification of religion is accentuated when the theory of evolution questions epistemology, metaphysics, bioethics, and theology.

By proposing that the "only other alternative [to evolution] is creation and this they [the evolutionists] all agree is untenable" (Nowers 1996). Staunch evolutionists are categorized as atheists. They never had the desire, the rationality, and the persevering will to seek the wonders of God's creation.

Evolution is rebellion against God. If evolution is correct there would be no sin. The nonexistence of sin would negate the need for a savior. Salvation would be unnecessary. In addition it would answer the query as to where the dinosaurs went. They simply evolved into another species. Evolution accentuates materialism in biology. It is a theory that eliminates an intelligent design in living organisms. But it does sustain a phylogenetic relationship between man and an anthropoid ape, the chimpanzee (pan troglodytes of western Africa).

Creation science is a misnomer in bioscience. Creation is not a scientific theory. Even the Supreme Court decided in the 1970s that Creation Science is religion. A meeting of the minds has no objection to teaching evolution as a theory. Like any theory there are *pro* and *con* discussions which are acceptable. Objection arises when evolution is not taught as a theory. When claims are made (without proof) that evolution is a historical fact then there is a parting of the way.

False knowledge becomes addictive and uncontrollable. Like the use of heroin it is non-alcoholic, dry drunkenness. Both lead to sin as does all untruths. Atheism and agnosticism are not divine blessings, without which the provocative power against sin is wanting. Unrestrained sin is malicious hostility to God. Sin is the blockade against the promise of eternal life. Like sin, evolutionary sentimentalists are decorously seductive, not dowdy, in the recruitment of candidates into their ranks.

Among the recruiting team was Thomas Henry Huxley (1825–1895), the nineteenth century's greatest popularizer of science. He was Darwin's perfect foil (Marschall 1998). Huxley attacked the Anglican Church persistently, acrimoniously, and mercilessly. His offensive avalanche of malicious criticism was not bermed by defensive contra-critics against his accusations. He returned to England in October 1850 after an extensive tour of duty as surgeon's mate on the frigate HMS *Rattlesnake* stamped him with an unyielding determination to enter the exclusive salon of Victorian science (Desmond 1997).

By 1876 Huxley* sailed the Atlantic for New York. He was now a successful scientist, well-known, and the world's first-class celebrity. T. H. Huxley had found a prime place in the esoteric scientific establishment wherein he was a superstar. To millions of people in the late nineteenth century Huxley was the epitomization of the major concept of an incarnate model scientist. His public lectures attracted all classes of Londoners who filled the seats to capacity as he described bones and oddities sent to him by far-flung naturalists who supported the puzzle of evolution. Their archeological findings were sent to him as ammunition against the doubters of Darwinism. During his lectures he always had time to blast old-fashioned clergymen, fusty professors, other academic stalwarts, and any person of distinction who opposed him and his philosophy.

Wherever he went, Huxley was uncritically lauded with general popular acclaim by similarly oriented non-religious fellow bioscientists. Their concurrence was an overall revelation of diffuse agnosticism in certain comparative circles of scholars. Conflicting opinions were not erased by the expansive reception of Darwinism. In its incipiency, evolution had an excellent showcase *via* the lively personality of Huxley. He fascinated those who listened to him, especially during the pivotal period of evolution's historical development. His facile flowing words needed no definitions to be understood.

As an iconoclast of divinity he had an innovative method of intertwining it surreptitiously and unoffensively while presenting the subject and allied matters at hand. Covering the topic with professional deportment, his descriptive explanations of the unverified, incomplete theory were superb. By his talent Huxley, more than any other evolution advocate, was able to bring an elucidation of the theory even to those who were divided by race, ethnicity, gender, ideological convictions, and special motivating interests. Using his dexterous, manipulative, theatrical ingenuity, Huxley elevated Darwinism to a fabulous, unique, cultural phenomenon.

As a sensationalist he was a recognized success, which was a vibrant tribute to his luminous, psychological power of impressionable persuasion. From a calm, reserved mood he could rise instantaneously to an unleashed affirming emotionalism. Modesty was not his brightest attribute. He was unfamiliar with the biblical quotation, "The greater thou art, the more humble thyself" (Ecclesiastics 3:18).

The end-result persisted when the jubilation ceased. Namely that evolution continued to be a contested, non-spiritual, unauthenticated theory minus accuracy of deduction and not upheld by fearless honesty.

In addition to his scientist label, taut Huxley acted like a philosopher, king of the domain or tyrant, depending on the analytical conclusion drawn. Insatiable for acclaim by an adoring public, he easily induced himself to ride roughshod repellently over established natural law, customary knowledge, and accepted customs. A morally dangerous addition to his appeal was that he was telling his listeners what they wanted to hear against religion. His magnetizing presence released science's indebtedness to the religious past as no longer an impediment mortgage to self-expression without restraint and no concern for God, ethics, and moral research restrictions.

As offshoots of his teaching and based upon some of his nonacademic lectures during his many scholarly peregrinations, Huxley wrote several textbooks. They contained lessons on elementary physiology which in essence were discussions on the general structure and functions of the human body. His writings placed much emphasis on the nervous system, especially the receptive organs of sensation. Secondarily he detailed neuromuscular movement and locomotion (Huxley 1869). Without arousing suspicion he drifted into evolution with laudation, *sans* locution, of its biologically implied correctness and his credence thereof.

Like a medieval mercenary, he was a freelance scientist. Although independent in speech, he was a forceful, committed companion to Darwinism. By strong contrast, Huxley underscored with enhancement the distinctive characteristics of Charles

Darwin. The former was a showman, ambitious, an extroverted commoner. The latter was a reclusive Brahmin, descendant of two generations of respected patriarchy, whose evolutionary publicity was extraordinary. Together they generated the dynamic energy that invigorated Victorian intellectual life (Desmond 1997).

"Huxley's attacks on the Church of England and on Oxbridge academicians were directed as much against inherited privilege as entrenched stupidity" (Marschall 1998). Whether Huxley was motivated primarily by his adherence to Darwinism or was the instigating stimulus to atheism *sui generis* is not judicable. Darwin may have lit the flame but Huxley blew the anti-religion flicker into a conflagration. Evolution and atheism were intertwined, becoming germane to each other.

By many ways, manners, and means, evolution is inhospitable to traditional Judeo-Christian, draconian dogma. Notwithstanding this religious non-receptivity, at times with drastic discussions, Darwinism has many worldwide adherents. Some of them are educated scientists who dismiss religion as an irrelevant nonentity in the race for the verisimilitude of Darwinism. Supported by bio-scientific *confreres*, a diverse group therefrom have manufactured scholastic pedestals to uphold the theory of evolution. This is an attempt by doctrinarians to elevate Darwinism to the status of a biological doctrine. Several proposals seem to have been begotten by visionary activists.

The repetitious writings, lectures, and college courses given on evolution keep it in existence, maintaining and prolonging its stature in bioscience. As the twentieth century came to a close, crusading cadres of bioscientists continued the recruitment of vulnerable tyros. Predictably, debate, discourse, dispute, and argumentation continue during the overture months of the third millennium and will for decades thereafter.

The initial years of the 1920 decade commenced with bitter disagreements among the rank and file in the Baptist Church. Loud oral and acrid written clashes were continual disturbances.

The plenteous southern fundamentalists and less numerous northern liberals battled over the proper interpretation of the Bible. One section believed in literal meaning. Opponents selected formal wide-field freedom in the readings. Prolonged debate culminated in the 1925 heated Scopes monkey trial.

Throwing off his former Baptist diffidence, John D. Rockefeller, Jr. inveighed against the "narrow and medieval creed" of his co-religionists, the fundamentalists. His far-ranged accusation was that they bred enmity and division. As a nationally known and philanthropic Baptist, his criticism received abundant attention. Never before had the junior Rockefeller expressed such adverse words against his church. With self-confidence, in the mid-1920s, he doubted openly the literal interpretation of the Bible. Believing strongly that the Scriptures were incompatible with modern sciences, he attracted many supporters. On this point he found agreement from his father who joined his son to figurative interpretation of the written words. "For fundamentalists, such heretical views diluted religion to a watery form of a social state. In 1926, with a mounting reaction, the Southern Baptist Convention reaffirmed the Genesis account of creation and unequivocally rejected the theory of evolution" (Chernow 1998). The quantity of northern Baptists may have been diminished by this declaration, but their denial of Darwinism persists.

Chapter 21

Pope John Paul II and Evolution

When the Holy Father addressed the Pontifical Academy of Sciences on October 23, 1996, he was not mantic to the stir that resulted therefrom. His intention was to encourage scientists in general to pursue dialogue between science and religion, as he believed fervently that scientific activity clarifies vision of the human person. Specifically, John Paul II's purpose was to extend his best wishes to the academicians on the occasion of their plenary session, which was on the sixtieth anniversary of the Academy's refoundation by Pius XI.* The plangent reverberation of his words was startling.

Misinterpretation of his French discourse may have resulted from its translation. This perhaps was the contributing factor to the disconnection of thought in its discretionary devolution. The Holy Father had established his basic philosophy in his previous dissertations which were repeated to the border of echolalia. The misguiding error could have been avoided, if the critics would have remembered that Pope John Paul II always reminded the world that scientific endeavors must always be grounded in truth. That truth must "shine forth in all the works of the Creator, and in a special way in mankind, created in the image and likeness of God" (Encyclical Letter *Veritatis Splendor* in its introduction).

This fundamental tenet was not altered in the Pope's address to the Pontifical Academy of Sciences. His Holiness is quoted:

I am delighted with the first theme which you have chosen: the origin of life and evolution—an essential theme of lively interest to the Church, since Revelation contains some of its own teachings concerning the nature and origins of man. How should the conclusions reached by the diverse scientific disciplines be brought together with those contained in the message of Revelation? And if at first glance these views seem to clash with each other, where should we look for a solution? We know that truth cannot contradict the truth (Leo XIII, Providentissimus Deus). However, in order better to understand historical reality, your research into the relationships between the Church and the scientific community between the sixteenth and eighteenth centuries will have a great deal of importance.

In the course of this plenary session, you will be undertaking a "reflection on science in the shadow of the third millennium," and beginning to determine the principal problems which the sciences face, which have an influence on the future of humanity. By your efforts, you will mark out the path toward solutions which will benefit all of the human community. In the domain of nature, both living and inanimate, the evolution of science and its applications gives rise to new inquiries. The Church will be better able to expand her work insofar as we understand the essential aspects of these new developments. Thus, following her specific mission, the Church will be able to offer the criteria by which we may discern the moral behavior to which all men are called, in view of their integral salvation.

Before offering a few more specific reflections on the theme of the origin of life and evolution, I would remind you that the magisterium of the Church has already made some pronouncements on these matters, within her own proper sphere of competence. I will cite two such interventions here.

In his encyclical Humani Generis (1950), my predecessor Pius XII has already affirmed that there is no conflict between evolution and the doctrine of the faith regarding man and his vocation, provided that we do not lose sight of certain fixed points. For my part, when I received the participants in the plenary assembly of your Academy on October 31, 1992, I used the occasion—and the example of Galileo—to draw attention to the necessity of using a rigorous hermeneutical approach in seeking a concrete interpretation of the inspired texts. It is important to set proper limits to the understanding of Scripture, excluding any unseasonable interpretations which would make it mean something which it is not intended to mean. In order to mark out the limits of their own proper fields, theologians and those working on the exegesis of the Scripture need to be well informed regarding the results of the latest scientific research.

Taking into account the scientific research of the era, and also the proper requirements of theology, the encyclical Humani Generis treated the doctrine of evolutionism as a serious hypothesis, worthy of investigation and serious study, alongside the opposite hypothesis. Pius XII added two methodological conditions for this study: one could not adopt this opinion as if it were a certain and demonstrable doctrine, and one could not totally set aside the teaching Revelation on the relevant questions. He also set out the conditions on which this opinion would be compatible with the Christian faith—a point to which I shall return.

Today, more than a half-century after the appearance of that encyclical, some new findings lead us toward the recognition of more than one hypothesis within the theory of evolution. In fact it is remarkable that this theory has had progressively greater influence on the spirit of researchers, following a series of discoveries in different scholarly disciplines. The convergence in the results of these independent studies—which was neither planned nor sought—constitutes in itself a significant argument in favor of the theory.

What is the significance of a theory such as this one? To open this question is to enter into the field of epistemology. A theory is a meta-scientific elaboration, which is distinct from, but in harmony with, the results of observation. With the help of such a theory a group of data and independent facts can be related to one another and interpreted in one comprehensive explanation. The theory proves its validity by the measure to which it can be verified. It is constantly being tested against the facts; when it can no longer explain these facts, it shows its limits and its lack of usefulness, and it must be revised. Moreover, the elaboration of a theory such as that of evolution, while obedient to the need for consistency with the observed data, must also invoke importing some ideas from the philosophy of nature.

And to tell the truth, rather than speaking about the theory of evolution, it is more accurate to speak of the theories of evolution. The use of the plural is required here—in part because of the diversity of explanations regarding the mechanism of evolution, and in part because of the diversity of philosophies involved. There are materialist and reductionist theories, as well as spiritualist theories. Here the final judgment is within the competence of philosophy and, beyond that, of theology.

The magisterium of the Church takes a direct interest in the question of evolution, because it touches on the conception of man, whom Revelation tells us is created in the image and likeness of God. The conciliar constitution Gaudium et Spes has given us a magnificent exposition of this doctrine, which is one of the essential elements of Christian thought. The Council recalled that "man is the only creature on earth that God wanted for its own sake." In other words, the human person cannot be subordinated as a means to an end, or as an instrument of either the species or the society; he has a value of his own. He is a person. By this intelligence and his will, he is capable of entering into relationship, of communion, of solidarity, of the gift of himself to others like himself. Saint Thomas observed that man's resemblance to God resides especially in his speculative intellect, because his relationship with the object of his knowledge is like God's relationship with his creation. (Summa Theologica I-II, q 3, a 5, ad 1). But even beyond that, man is called to enter into a loving relationship with God himself, a relationship which will find its full expression at the end of time, in eternity. Within the mystery of the risen Christ the full grandeur of this vocation is revealed to us. (Gaudium et Spes, 22). It is by virtue of his eternal soul that the whole person, including his body, possesses such great dignity. Pius XII underlined the essential point: if the origin of the human body comes through living matter which existed previously, the spiritual soul is created directly by God (animas enim a Deo immediate creari catholica fides non retimere iubet) (Humani Generis).

As a result, the theories of evolution which, because of the philosophies which inspire them, regard the spirit either as emerging from the forces of living matter, or as a simple epiphenomenon of that matter, are incompatible with the truth about man. They are therefore unable to serve as the basis for the dignity of the human person." (America Online: Gianicolo Pages 1, 2, 3).

In its November 14, 1996 edition, *Origins Magazine* published a translation of Pope John Paul II's French language message to the Pontifical Academy of Sciences. Based on this translation, it was printed in the English issue of *L'Osservatore Romano*, the Holy See's official newspaper. In that issue the editor points out a discrepancy with an explanation to correct the difference in meaning. The revised text highlights the passage when the Pope states that over the past fifty years, new knowledge has emerged showing the theory of evolution to be "more than a hypothesis." The original French is, *"de nouvelles connaissances conduisent a reconnaitre dans la theorie de l'evolution plus qu'une hypothese."* Included in the papal message was that "Revelation contains teaching concerning the nature and origins of man."*

Thus the Pope in making the above statement riled fundamentalists and literalist interpreters of Holy Scripture. God is the why of creation. Evolution is a proposal of how it came about. Pope John Paul II is not the first pontiff to consider this theory. In his encyclical *(Humani Generis* 1950) Pope Pius XII expressed almost the same thought. This writing disturbed many people even as the words spoken by this living Pope aroused discord by his address to the Pontifical Academy of Sciences on October 23, 1996.

When Charles Darwin published his revolutionary text *On the Origin of Species by Means of Natural Selection* in 1859, the book was opposed with vigor by many Christian denominations. Pope Pius IX* never made any official condemnation of the Darwinian document. His persistent silence was not broken when strong advocates of Darwinism attained great heights of authority. Even the sustaining lectures of T. H. Huxley and Herbert Spencer, whose influence from England invaded continental nations, did not cause Pope Pius IX to depart from his silence. Militant materialists, intellectual atheists, and popular agnostics did not move Pope Pius IX to express any words on evolution.

It becomes evident that precise reading of Pope John Paul II's message to the Pontifical Academy is not a novel proposition on evolution. Rather it is a restatement on the theory expressed by Pope Pius XII. This was contained in his encyclical *Humani Generis.* Pope Pius XII considered Darwinian materialism the model for communism. (Evolution was embraced enthusiastically by Karl Marx.) The only new element is the acknowledgment that the theory of evolution, which for Pius XII had been only an hypothesis, is worthy of further research and reflection (along with research and consideration of opposing theories). As John Paul II reflected on creation and evolution, the mass media distorted his meaning. "The present Pope does not affirm that evolution has become a certain demonstrable doctrine. In the Holy Father's own words: Rather than speaking of the theory of evolution, we should speak of (various) theories of evolution—since there does not seem to be a unanimity among scientists" (Gaspari 1997).

The Vatican specialist on anthropological studies, Reverend Vittorio Marcozzi, was an advisor to three Popes. In 1997 he was eighty-eight years old and was considered an authority on evolution. Hence he was summoned to the Congregation of the Doctrine of the Faith to debate with eminent, worldwide scientists who had definite opinions on evolution and creation. His initial statement was that scientists should not discuss the subject under the title of creation versus evolution. Preferably the subject should be creation and evolution. Furthermore, he maintained that to admit evolution does not necessarily signify denying God's intervention.

Father Marcozzi stated in his discussion that there are at least three moments when divine intervention is necessary and evident. These are at the appearance of life (that is of the first living organisms); the evolutionary possibilities with which God imbues these organisms; and finally, the coming of mankind, whose spiritual qualities implicate God's special intervention (Gaspari 1997). Other pivotal statements made by him are litanized for brevity (Marcozzi 1997).

1. *Evolution is not admissible without the mediation of a supreme mind which established the laws of nature governing natural processes and which created nature itself.*
2. *Darwin was a materialist criticized by his own wife for his lack of faith.*
3. *Evolution was set in motion by outside causal factors such as natural selection and the struggle for survival according to*

Darwin who believed that all beings, including man, evolve from causal mutations.

4. *Apart from the absence of clear proofs for the intermediary forms of human existence, is it plausible that such marvelous beings, particularly human beings, are products of mere chance?*
5. *A billion and a half years have passed between the existence of one-celled and many-celled organisms, yet there seems to be no intermediate forms linking the two.*
6. *God created all things: evolution in no way contradicts this affirmation.*
7. *In synthesis, God created human beings from matter and then infused them with spirits.*

When asked his opinion of the Holy Father's message to the Pontifical Academy of Sciences, Marcozzi responded clearly. "The Holy Father's message contains no specific recognition of Darwin or his theories. John Paul II is proceeding along the doctrinal lines traced by the Popes before him. There are many different theories of evolution. It is possible to accept evolution as a theory, while affirming that the spiritual and philosophical elements must remain outside the competence of science" (Marcozzi quoted by Antonio Gaspari 1997).

Other quotable sentences worthy of recording on evolution are listed herewith.

1. *New knowledge leads to the conclusion that recognition must be given to the concept that there is more than one hypothesis in the theory of evolution.*
2. *If the human body takes its origin from preexistent living matter, the spiritual soul is created immediately by God.*
3. *Darwin's theory of evolution contains elements in obvious contradiction with Christian faith: in addition to contradicting Bible accounts of creation, it suggests animal origins for human beings.*

4. *The development of the spiritual aspect of human life cannot be explained scientifically.*
5. *Science deals with the measurable aspects of reality: religion deals with the spirit. There is no conflict between them for each seeks the ultimate criterion of truth.*

The Holy Father calls attention to truth. One of his magnificent encyclicals glorifies the splendor of truth. With this emphasis on truth, John Paul II, as an extremely well-read, learned, intellectual scholar is aware of valid, scientific research. Hence in his discussion on evolution before the Pontifical Academy of Sciences, he affirms unequivocally that divine intervention created mankind at a certain point in natural history, conferring upon the human species a particular physical and metaphysical character.

Mankind has been and continues to be called upon to enter into a relationship of love and knowledge with God. This relationship finds its complete fulfillment beyond time—in eternity. All the solemnity, depth, beauty, and incomparable grandeur are revealed to those who have the grace to hear, listen, and receive it in the mystery of the risen Christ (cf. *Gaudium et spes,* N22). It is by virtue of his spiritual soul that the whole person possesses such a dignity even in his body (Pope John Paul II 1996).

In the domain of inanimate and animate nature, the Darwinian evolution and the evolution of science with their applications give rise to new questions. The better the Catholic Church's knowledge is of these essential aspects, the more she will be able to evaluate them. Through divine providence, diligent care, intellectual preparation, and advanced foresight through spiritual direction, proper conclusions will be arrived at based upon transcendental truth because truth can never contradict truth (cf. Leo XIII encyclical *Providentissimus Deus*).

Evolutionary scientific creationism tweaks the Christian establishment and irks the fundamentalists. This is not a new situation. Religion and science have been enemies for many centuries.

Always they stressed their differences, rarely seeking a mutual meeting of thoughts to clarify disparities. Unreflective scientism seems to rejoice when it contributes to the weakening of the foundations of religious faith and its practice. When attempts are made for mutual understanding, haphazard desultory dialogue is the result. One of the identifying characteristics of Pope John Paul II's pontificate has been his friendship with scientists and open-mindedness to science. He has been the seeker of harmony between science and religion. The Pope strives for reconciliation of science with faith, philosophy, ethics, and religious attitudes. Aggressive secular scientific organizations and individual agnostic groups undermine many attempts made to establish harmony between religion and science. Even if discussions occur the religious viewpoint does not prevail.

When scientists meet with religionists, they enter dialogues at antipodes with each other after the first sentence is spoken. Any *agora*[*] does not alter this tendency. Obvious from the start is the current bias in the scientific community that reflects outright denial of religious truth or has moderate opinions that relativize[**] it. Hence it is demonstrated quite forcibly that "the content and method of religious truth-seeking are radically different from scientific content and method" (Byers 1996).

Science focuses on the quantifiable aspects of things apparent to the aided and/or the unaided senses. Relationship to God is omitted. No consideration is given to salvation, why are we on earth, the purpose of human life, and what is humanity's final destiny. Scientists are inarticulate on spirituality. Outside their own specialties they have little interest in matters supernatural. Visible phenomena may captivate them but noumena has no charm, elicits no curiosity, and is discounted. Hence differing epistemologies disenchant a satisfactory dialogue between them with mutual respect and equal solemnity. The fundamental barrier that separates science from religion is faith that finds its origin in revelation, tradition, and the life of Jesus Christ.

"Science can purify religion from error and superstition; religion can purify science from idolatry and false absolutes. Each can draw the other into a wider world, a world in which both can flourish" (John Paul II).

Chapter 22

Exaggerated Darwinism

Like many popular theories, proposals, and notorious concepts Darwinism exerted an influential persuasiveness in other areas. One discipline can and often does stimulate postulates in other unallied subjects for speculation. Hypothetically fascinating to diverse researchers, the popularity of evolution carried the theory into the realms of sociology, economics, genetics, epidemiology, and other undisclosed disciplines.

After the promulgation of Darwinism, English sociologists and economists maintained that the plight of the underclasses was due to their failure to survive in the world of competition. Meaning they were not able to overcome whatever was deterring them from successful accomplishments. Adherents to this belief ignored other factors that produced this socioeconomic disability. Total blame was thrust on the person. No concern was given to the contributing socioeconomic conditions depicted in the masterful novels of Charles Dickens.

In the last decade of the twentieth century newer applications of evolution were presented to the scientific community. In a publication, *The Evolution of Growth and Development,* the author "explores evolution down a much neglected avenue that links natural selection and genetics, the effect of changes to the rates and timing of growth and development. [The author] delves into the living and fossil worlds to show how animals and plants have evolved when the carefully orchestrated pattern of embryological development is gently nudged off-course. [The writing] [s]hows how this phenomenon known as heterochrony has affected many aspects of evolution, including the mechanism behind the selection of different breeds of animals, differences between sexes, and animal behavior. Heterochrony explains how the dinosaurs got so big, how pterosaurs managed to produce a wing supported only by their fourth fingers, and what has driven the primate species with the biggest brain and longest childhood, *Homo Sapiens*" (McNamara 1997).

A textbook, *Tropical Veterinary Medicines,* "examines the molecular tools used in epidemiology to define the evolutionary history of organisms and their relationship to other organisms." (Jongejan, Goff, and Camus 1998). Utilizing the concept of evolution as the basis for unfolding molecular epidemiology of hemoparasites their vectors are analyzed. In this project the researchers look for differences between attenuated vaccine strains and virulent field strains and study the dynamics of disease transmission in a selected population.

An appealing reference session centered on "Molecular Strategies in Biological Evolution." It stressed the lineages of organisms' response to challenges and opportunities in their environment. Exploration was made of the notion that organisms have evolved the ability to modulate the rate, location, and extent of genetic variation. "Jumps in efficiency, made possible by development of novel efficient evolutionary strategies, could fuel rapid, salutary expansion of species into novel niches as each innovation evolves. An up-to-date assessment is provided on biochemical mechanisms available to modulate the rate of genetic change at specific sites within a genome" (Caporale 1999).

Invasion of evolutionists into the staid academic corridors of science found resistance by opponents to this theory. But no one tried to silence Darwinian scientists. In the declining decades of the twentieth century many efforts by individuals, groups, and formal organizations tried to hinder scientific progress, "using techniques ranging from defunding scientific research to quashing scientific debate to substituting junk science for real science" (Milloy and Gough 1999).

"It's a good thing junk science is only a relatively recent phenomenon. In American Lawyerland, pseudo-science would have deemed microwave ovens, TVs or medical X-rays dangerous cancer risks from which only the legal industry could save us. Junk science players win by convincing us that our nation is in peril from its scientific progress. But victims of the *Escherichia coli* bacterium did not win when meat irradiation approval was delayed" (Popeo 1999).

Worthy of mention is the adoption by philosophers of a scientific principle as a contribution to their discipline. An illustrative example was Einstein's relativity. By some sophistry, many philosophers believed everything was relative in the physical, metaphysical, and moral order. Relativism evolved into subjectivism that was used to justify innovative thoughts, words, actions, and immoral performances. What commenced as a novel philosophical attraction became an intellectual subtraction.

Promoters of Darwinism believed its future was bright initially. Soon after its promulgation clouds of controversy swirled around the theory. Not ignored by many critics was the commercialism attached to its participating advocates. Others sensed it was an outrage against the strict ethics of formal science. Adverse criticism did not deactivate its progression.

Amid the surge for recognition the pro-evolution missionaries were determined to expand the open minds of undergraduate students by this new biology enlightenment. Simultaneously, efforts were made to lure mature scientists into its ranks. With much encouragement to accept the new theory, responses were an outburst of activism which spurred others to urge evolution's acceptance in bioscientific societies, intellectual circles, and at universities.

Disagreement with Darwinism was expressed by a large

number of intelligent scholars. Without preconceived prejudice they discerned no bioscientific proof in evolution. Its merit was not substantiated as factually correct. Demerits existed even when considered either as a hypothesis or a theory. Because there is lack of scientific credence, evolution is a belief. Atheistic evolution under the light of right reason is "false according to rigorous scientific methodology . . . the *illusion* of proof is facilitated by the way some evolutionists constantly shift ground between various conflicting, supposed mechanisms for evolution" (Keane 1999, 7). Protestation against it as unscientific and its condemnation as anti-biblical is apparently futile in an ambiance of science's agnosticism. In the archival field of biomedical history evolution has not been buried in the undergrowth of supposition and misbeliefs.

The evolutionary, anthropological, and archeological facets of Darwinism may be a fascinating intellectual exercise for the mind oriented to bioscience. However, the relevance to academic self-indulgence for the righteousness of the cause is not upheld beyond its status of an undeciphered theory. No introspective understanding exists to champion evolution as an emerging, shattering scientific dogma. Social historians may elect to categorize Darwinism as a viable theory. However, it is not legitimized as a discipline of "classic configuration."*

Under erudite scrutiny, meticulous insight, and strict scientific overview the concept of evolution does not emerge as a lucid, pioneer contribution to bioscience. Although it has done much to foment discord, discussion, and dialogue among bioscientists, sociologists, and historians; it has not brought about a more reflective and sophisticated elucidation with professional expertise of the subject that clarifies it more clearly than does the original publication.

Evolution remains a topic for discussion and discontent. Decorated in modern dress it is presented anew, bringing more chaos than orderliness to the existing confusion and precarious empiricism. Keen mental penetration into evolution displays flaws in rational deduction and other extraordinary capacities. In their zeal, many Darwinians defied reason. Perhaps they failed to read "The two chief Pillars of Physick are Reason and Observation: But Observation is the Thread to which Reason must point" (Baglivi 1704).

As a new theory, evolution was neither innovative nor sublime. Although considered by its backers as magnificent, the convulsive gasping disenchanted religionists. Disquietude was intensified by the template song sung by many who were thoroughgoing materialistic opportunists. Turning a deaf ear on the divisive disputes, there was no tranquility of mind among the chief proponents of evolution who were lukewarm to theology. With all the accolades heaped upon Darwin, no positive proof is given that he was a judicious thinker and an original genius. His writings are not replete with reason but are crowded with observations. The scientific fragility of Darwinian evolution has not caused it to vanish.

Epilogue

At the beginning of the nineteenth century Lamarck* tried to explain the development of organisms from the simplest to the most complex. This included human beings among the complicated. It was not until Charles Darwin came on the biological scene that naturalists were exposed to the theoretical mechanism of the origin of species. Darwin made them aware that natural selection operates on the basis of hereditary variability, admitting that the laws governing inheritance were unknown, demonstrating his lack of knowledge of Mendelism.

Darwinism and the application of survival of the fittest was postulated in disciplines allied to and in others alien to biology, even becoming a focusing influence on some contemporary natural philosophers. During the past one hundred forty years plus since Darwin put forth his theory, different scientists have penetrated into his beliefs as they investigated all levels of living biological systems.

In the last decades of the twentieth century many studies, speculations, and suppositions have been published on evolution. Historians, philosophers, theologians, doctoral candidates, sociologists, and bioscientists have a profound curiosity of what is learnable about evolution. As is anticipated, opinions differ in the evaluation of the various interpretive aspects of evolution and its scientific significance. Additionally, many investigators are fascinated by and are eager to learn under what modest conditions a scientist who can be called an innovator in the life sciences arrived at such a dramatic, thought-provoking, controversial theory.

Darwinism emerged as more than a small transient riffle over the rocky shoal in the fast-flowing stream of bioscience. Evolution had its *prima donna* role in the portrayal of life. Without cryptic, concealed advocates, words were spoken that jolted listeners with whiffs of both mystery and menace. No presenter was reticent in enthusiastic support of a specific, positive opinion. Neither defender nor detractor evoked in the other either an imposed silence or an instantaneous change of thought. None implied possession of omniscience but some did not always display humility.

Non-scientists have accepted the theory of natural selection as a conversational topic with reliant possibility based on what they term common sense. Meaning it is feasible, even though Darwin, his supporters, and emotional devotees have not given proof for its acceptance. Forceful arguments against Darwinism concern its failings as science, not its contradiction to religion. Darwin's faith was in his strong belief that evolution was real. Blind faith in his unproven doctrine of natural selection was greater than any belief in the Book of Revelation. Hence the antagonism between religion and evolution had an indigenous origin in its authorship.

Darwin was so certain of God's nonexistence that he ignored any evidential proof against his theory after it was postulated in print. Die-hardened materialism drove Darwin's mind, not scientific research, and not verifiable, scientific observation. Although many world-famous, major scientists have dissented, their objections have been drowned by the waves of atheism that have pervaded the intellectual community for the last one hundred fifty years. Ideology rather than science supports evolution. Empirical science has the defining mark of falsifiability that is demonstrable by scrutiny. Evolution cannot be affirmed as true based upon observation, experimentation, or any other methodology that can be its vehicle of verification.

Prudery was not demonstrated by religionist critics. But they did not hesitate to disparage the evolutionary theory as diminishing common religious beliefs and the Bible as adscititious. Darwinism admittedly had jostled the intelligentsia. Academia was not slow to react. Sometimes with overreaction but never with underreaction after an in-depth assay of the revenant subject with its exacerbations, dormancy, and reactivated recrudescences. On both sides of this almost calamitous dispute were more questions than answers. Some opinions were expressed without meekness although apparent sympathy for them was engaged. The disputations to this day have not diminished the diversity of interpretations on the drama of evolution which in some religious circles and bioscientific quarters has been intense.

A disproportionate number of scientists who are underqualified and underprepared in theology speak/write on this subject without presenting well-authenticated, adequate reasons for negative expressions against the existence of God. Appended to which is the categorical denial of creation. Knowledge of philosophy is conspicuous by its absence. Non-syllogistic arguments are faulty in their incipiency without any rectification. Thus they are effective temporarily, but lack permanent stability of thought.

A question has been asked. Are the negative notions of divinity simply a matter of Darwinian convenience? (Kafka 1998). Current studies in neuroscience are the number one attraction on university campuses worldwide. An impressive neurological researcher claims he has found specialized brain circuity in the temporal lobe that may make people more prone to believe in God (Ramachandran 1998). The basis for this conclusion resulted from tests on patients with epilepsy. These proven epileptics showed that words relating to religion induced a measurable galvanic response. The observed recording corresponded to an electrochemical reaction in the brain not replicated by other stimuli of equal and greater intensity such as violence (Kafka 1998).

If it is true that the brain has a built-in "God module," as Vilayanur Ramachandran theorizes, does that mean adverse notions of divinity are simply a matter of Darwinian convenience? This question is raised by Peter Kafka. The relationship of Darwinism as a biological handy suitability and divinity as a theological certitude recognized by the temporal lobe of the brain is not clarified. Darwinism inserted into neuroscience has an ambiguity that remains a conundrum encased in a circle of curiosity.

Although the brain circuity in the temporal lobe has no bearing on whether God exists or does not exist, no volunteers are found who are willing to discuss the subject. Thus this neuroscientific research seems to touch upon the dictum of François Marie Arouet Voltaire (1694–1778) who wrote more than two centuries ago, "If God did not exist, it would be necessary to invent him." More modern language would be, "If God does not exist, let us bioengineer Him." An easier solution to the indecisive quagmire of this frustration is God exists and perhaps placed the brain circuitry in the temporal lobe (Kafka 1998).

Delving into the depths of brain function and the mechanical electrical tracings made by external variegated stimuli do not solve the mystery of faith in God. The greater the efforts expended, the less the scientists seem to learn about God. Without faith the old divine mysteries remain mysterious.

Darwin was a reversed, retrograde clairvoyant as he looked backward on the origin of species. He possessed no cryptesthesia as to the slathered agnosticism he was insinuating into bioscientific thought. Assimilated information energizes mental power to seek more knowledge. Danger lurks nearby when in the quest for learning no differentiation is made between truth and speculation.

Persons dedicated to the acquisition of scholarship should immunize themselves against the three comorbidities of science. These are dishonesty, failure to recognize truth, and trumped up experimentation. Without stabilizers scientists have no deterrent to prevent them from going in a rakish, wrong direction. They become entangled in convoluted actions which are radical if not tragically dethroning.

Disparagement of evolution has been attempted. It has been ineffectual even when swathed in anti-religious insinuations. Stentorian words spoken and those written have not destroyed Darwinism. No diminishing enthusiasm for the theory is tangible. It flourishes among many bioscientists and non-scientists without any permutation of its originality.

The strength of evolution is manifested by the softening of voices in the creation-evolution debates that continue. A new tendency is discerned. Scientists are acknowledging limitations. Creationists are admitting extremes in their criticism and crusades against evolution. In a 1996 debate at Calvin College, Grand Rapids, Michigan, a leading evolutionist was matched with an arch critic of Darwinism. "Taking the debate to this Northern setting, a Christian college founded by Dutch Calvinists, illustrated that the new level of argument is not the Bible versus science, but naturalism versus theism . . . One of the world's leading evolutionary theorists* insisted that there is overwhelming evidence of human descent from apes and it need not refute anyone's faith" (Witham 1996).

Contrarily a professor stated his view that naturalism—only material or physical evidence gives human beings real knowledge—is a dogma of higher learning that may be on its last legs.* "The materialism of Karl Marx and Sigmund Freud already has fallen . . . but Darwin still retains a tremendous authority and prestige . . . There is remarkably modest scientific evidence that evolution created complex beings . . . The demonstrations of the power of natural selection are very modest" (Johnson 1996).

"The topic of evolution long has been known for its great debates. The first was between Darwin's most aggressive defender, Thomas H. Huxley, and the Anglican Bishop Samuel Wilberforce at Oxford University in 1860. Huxley portrayed the exchange as the intellectual defeat of creationism" (Witham 1996).

The dramatic oral exchanges between the Episcopal priest and the ebullient, effervescent demagogue of science was a colossal spectacle. Evolution came to life. The audience responses to the theory demonstrated absorption of the words spoken with much alacrity and celerity. Huxley conveyed his message crowning evolution in the setting of a sparkling gem. The clerical garbed bishop preached as if to his congregation that evolution was an effrontery to the Divine Creator. Cloaked with professional dignity, the debate was balanced extraordinarily between the contestants without a terminal victory distributed equally. Although extremely insightful, beautifully presented, and impressively biblical, Bishop Wilberforce was vanquished.

The scene was an exhilarating mental journey through the remote corners of global evolution. Both speakers pondered with tense momentum the pro-scientific and the anti-religious inferences of Darwinism. The scientist confronted the clergyman with tolerance but disclosed incredulity. Without any apparent indication of prejudice the bishop displayed disagreement. But the clergyman had neither malice nor hatred for his opponent.

The conterminous contestants excelled in the scholarship of their respective, chosen field of professional endeavor. Each had an in-depth, wide breadth of knowledge which could easily influence scholars in other masterful disciplines. In turn they produced mild to moderate effects in the public perception of fundamental religions and scientific ideas. Their energy fueled discourses that could shape the thinking processes of those who listened attentively. With remarkable Demosthenes-like persuasiveness, each stamped a mark on venues outside their own career realms. Since the landmark debate of 1860 the dissonance between religion and evolution continues unabatedly down to these current years beyond the birth of the third millennium.

As the radiant rays of the new century are burgeoning, the creation-evolution debate may become more moderated. The shrill tone of discord seems to be lessening. Discordant cacophony that drowned out comprehension may be replaced with symphonic language of mutual cooperation in logical reasoning of disputed discourse. All participants should be more attentive and less disruptive with sudden outbursts of disagreement. Disputants should become discussants with honesty present openly and bilaterally. Caustic belligerency and non-subtle innuendo should be dismissed from all debates with their relegation to obscurity a mutual acceptance. Attraction to evolution is not deceased. Many of its advocates have acquired new recruits from academia and learned scientific groups. Those advocating the teaching of evolution in public schools have many followers who believe evolution to be "the most important concept to modern biology" (*New York Times* 1998).

The National Academy of Sciences published a guidebook for teachers and others which contains the statement that "There is no debate within the scientific community over whether evolution has occurred and there is no evidence that evolution has not occurred. [It also says that] understanding evolution is essential to understanding vital processes, like how bacteria become resistant to antibiotics . . . Many students receive little or no exposure to the most important concept in modern biology" (National Academy of Sciences 1998).The guidebook also raises these points to answer those who are in opposition to evolution:

1. *People can still believe in God and accept God because religion and science answer different questions.*
2. *Fewer than one-half of American adults believe human beings evolved from earlier species, and more than half want creationism taught, according to surveys.*
3. *Children should be graded on their understanding of evolution but not penalized for refusing to believe in it. It is quite possible to comprehend things that are not believed.*
4. *To set the record straight, human beings did not evolve from modern apes, but humans and modern apes shared a common ancestor, a species that no longer exists.*

A high school biology teacher in Wichita, Kansas, said he avoided talking about human descent from earlier primates and instead focused on genetics and change when talking about evolution. Regardless of other explanations of how life began, you have to understand it to understand biology and make sense out of it (Raugust 1998).

Calmness and uniformity of opinion do not prevail in the world of education. "Evolution still causes trouble for teachers and school officials more than seventy years after John Scopes was convicted of violating a Tennessee law against teaching it and more than a decade after the United States Supreme Court ruled that public schools cannot teach that God created the universe" *(New York Times* 1998).

Hopefully, prayerfully, in the third millennium and thereafter the distant relationship between science and religion will become closer as wisdom brings about a pleasant metanoia where needed. A favorable confluence of science and religion will become a complementarity from which can be emitted intellectual, moral, and beneficial productivity that promotes human munificence.

Sciences in general and bioscience in particular have become dynamic forces in social changes. Not as a policy establishment by activist scientists, but by dint of its impact on the progress science has added to the ongoing process of civilization. This is not applicable to the woeful state of the knowledge base for either sustaining the theory of evolution or strengthening it. The proof of its accuracy is slow in coming as the layers of doubt covering it have not been turned over satisfactorily to make it more comprehensible and unravel what is missing.

Stronger proven bonds are required between theorizing and proving the validity of evolution. Not extant is any intellectually acquired gain that outstrips the loss of credibility in evolution. Outputs of proof are not strong enough to diminish doubt and increase credulity. However, evolution's acceptance is not stagnant. Many professorial stalwarts in the academic world are supporters of it as they continue to be fascinated with its possibilities to explain what their curiosity stimulates.

Evolution remains an incomplete picture. To finish the undone painting, more than sluggish bioscientists are required. What is needed are those gifted with intellectual artfulness in humanity to harness the adverse criticisms and eliminate all doubt. By discovering and organizing the missing transitions in a convenient, understandable, verifiable, scientifically accurate spreadsheet, this can be accomplished. Only then will evolution be more than a moment in the vast, continuous, legendary life of nacreous science.

To many non-conformists to scientific principles, evolution achieved symbolic importance. Especially was it made mesmerizing by the dramatic presentations of it by paid lecturers and commercial entrepreneurs. All the persiflage was out of proportion to its actuality. Byzantine chicanery lured some bioscientists to become zealots in its behalf. Although the ridiculousness was discernible, it resulted in no loss of self-esteem, because the audience was dazed.

Readers did not analyze what was written and listeners did not hear the falsities. Often the lectures on evolution and the exhibits of accumulated specimens were looked upon as amusements. As an imprecise, fallible process with unsubstantiated assumptions, it passed undetected. The discouraging imperfections in Darwin's evolutionary theory has neither daunted its enthusiasts nor made them crestfallen.

Bibliography

Alter, S. *Darwinism and the Linguistic Image.* Baltimore: Johns Hopkins University Press, 1999.

Austad, S. Professor of Zoology, University of Idaho. *On Aging Process.* Quoted by Doug Stewart q v infra.

———. A research ornithologist at the University of Idaho, he is associated with Donna Holmes in a study of repair mechanisms that protect bird cells from oxidative damage and consequently slow down aging rate. Quoted by Montgomery q v infra.

Azzone, G. *Medicine from Art to Science: The Role of Complexity and Evolution.* In IVSLA series, vol 1. Amsterdam: IOS Press, 1998.

Baglivi, G. *The Practice of Physick, Reduced to the Ancient Way of Observations.* London: Printed for Andre Bell et al., 1704. Translation of Baglivi's *De Praxi Medica ad Priscamobservandi rationem revocando* printed in Rome by Typis Dominici Antonii Hercules, 1696.

Bateson, W. *Mendel's Principles of Heredity: A Defense.* London: Cambridge University Press, 1902.

Becker, D. "Creation or Evolution? A Call to Intellectual Conversion." *Homiletic and Pastoral Review* (April 1993): 54–61.

Behe, M. *Darwin's Black Box: The Biochemical Challenge to Evolution.* New York: Free Press, 1996.

Bergen, J. and F. *The Development Theory.* Boston: Lee & Shepard, 1884.

Bergson, H. *Creative Evolution.* New York: Holt, 1913.

Bibby, C. *Scientist Extraordinary. The Life and Scientific Work of Thomas Henry Huxley.* New York: St. Martin's Press, 1972.

Bishop, J. "Enemies of Promise," *Wilson Quarterly,* (Summer 1995): 61–64.

Bork, R. *Slouching Towards Gomorrah.* Quoted by Stephen Jay Gould q v infra.

Bower, B. "Objective Visions." *Science News,* Washington, D.C., (December 5, 1998). Quoted in *Wilson Quarterly*, vol. 23:2 (Spring 1999): 112.

Bowler, P. *Charles Darwin: The Man and His Influence.* New York: Cambridge University Press, 1996.

———. *The Eclipse of Darwinism.* Baltimore: Johns Hopkins University Press, 1992.

———. *Theories of Human Evolution, 1844–1944.* Baltimore, MD: Johns Hopkins University Press, 1986.

Burkett, W. *Creation of the Sacred Tracks of Biology in Early Religions.* Cambridge: Harvard University Press, 1996.

Burkhardt, F. *Charles Darwin's Letters: A Selection 1825–1859.* New York: Cambridge University Press, 1996.

Burkhardt, F., D. Porter, and H. Topsham Jr. *The Correspondence of Charles Darwin*. Vol 10, 1862. New York: Cambridge University Press, 1997.

Buss, D. *The Dangerous Passion.* New York: Free Press, 2000.

———. *The Evolution of Desire* published in 1994, quoted by Lee Allen Dugatkin, professor of biology at the University of Louisville. David M. Buss is professor of psychology at University of Texas.

Byers, D. "Religion and Science: The Emerging Dialogue." *America,* vol 174:3 (April 20, 1996): 8–15.

Cambridge Encyclopedia on Human Evolution. Delran: Library of Science 1998 (P.O. Box 6014, Delran, NJ 08075-9649).

Caporale, L., editor. "Molecular Strategies in Biological Evolution." Proceedings of a conference sponsored by the New York Academy of Sciences on June 27–29, 1998; vol. 870, 1999.

Caputo, J. *Deconstruction in a Nutshel.* New York: Fordham University Press, 1997.

Chernow, R. *Titan, the Life of John D. Rockefeller, Sr.* New York: Random House, 1998: 640.

Chesterton, G. *Darwinism.* Quotation from Roche q v infra.

Clark, H. *Mind In Nature; or, The Origin of life, and the mode of development of animals.* New York: Appleton, 1865.

Coleman, W. Ferdinand Schindler's Letters to William Bateson, 1902–1909. *Folia Mendeliana,* 2:9–15, no year.

Colp, R. Jr. *To Be an Invalid: The Illness of Charles Darwin.* Chicago: Chicago University Press, 1997.

Congdon, J. Head of research on Blanding turtles at Edwin S. George Reserve in southeastern Michigan over 45 years. Quoted by Montgomery q v infra.

Cravens, H. *The Triumph of Evolution. American Scientists and the Heredity-Environment Controversy 1900–1941.* Philadelphia: University of Pennsylvania Press, 1978.

Dalrymple, T. "My face or yours." *The Spectator*. London (May 22, 1999): 12–13.

Darwin, C. *Journal of Researches into the Natural History and Geology of the Countries Visited during the Voyage of H.M.S.* Beagle *round the World.* New York: D. Appleton and Co.,1880.

———. *The Descent of Man and Selection in Relation to Sex,* 2nd ed. London, 1877.

———. De l'origine des espèces par sèlection naturelle ou des lois de transformation des êtres organisés. Traduction de Mme Clémence Royer avec préfaces et notes du traducteur. Lyons: P. Guillaumin, 1870.

———. L'origine des espèces au moyen de la sélection naturelle ou la lutte pour l'existence dans la nature, traduit sur l'édition anglaise définitive par Ed. Barbier. Paris: P. Reinwald, 1880.

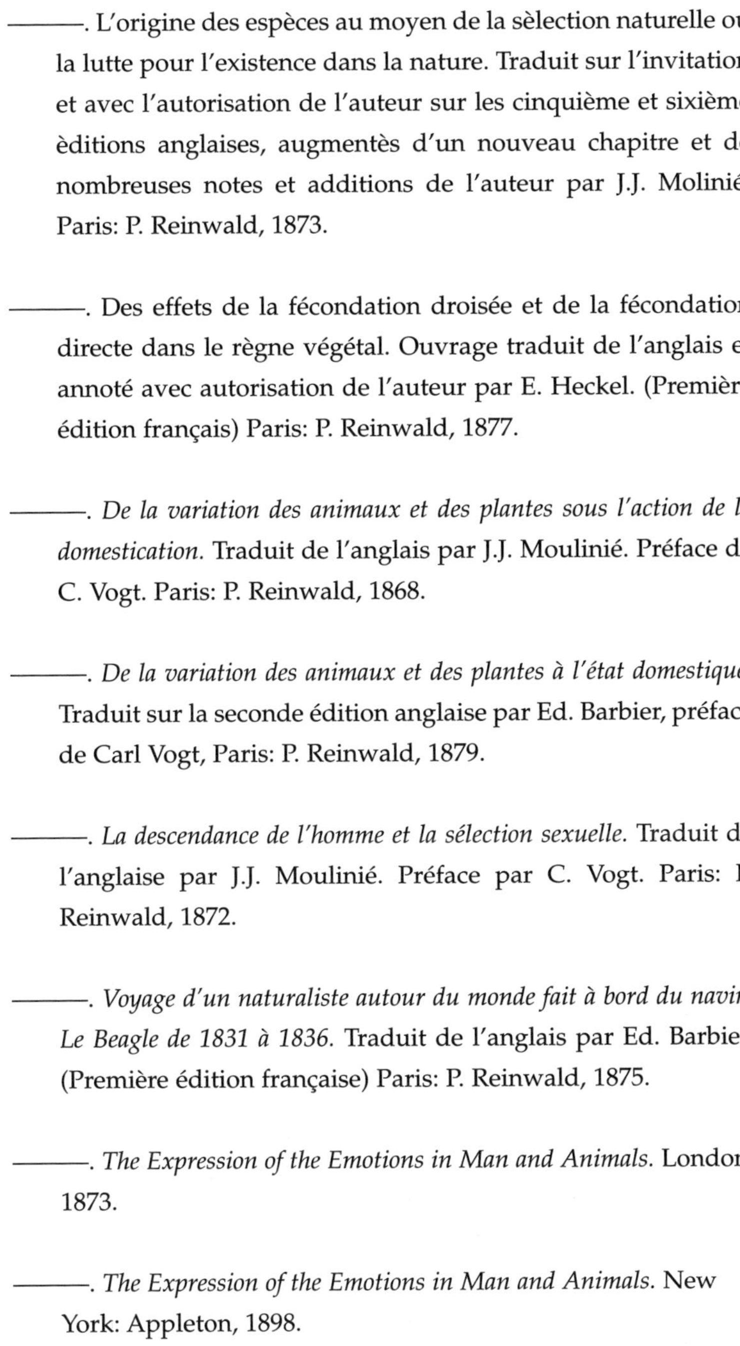

———. L'origine des espèces au moyen de la sèlection naturelle ou la lutte pour l'existence dans la nature. Traduit sur l'invitation et avec l'autorisation de l'auteur sur les cinquième et sixième èditions anglaises, augmentès d'un nouveau chapitre et de nombreuses notes et additions de l'auteur par J.J. Molinié. Paris: P. Reinwald, 1873.

———. Des effets de la fécondation droisée et de la fécondation directe dans le règne végétal. Ouvrage traduit de l'anglais et annoté avec autorisation de l'auteur par E. Heckel. (Première édition français) Paris: P. Reinwald, 1877.

———. *De la variation des animaux et des plantes sous l'action de la domestication.* Traduit de l'anglais par J.J. Moulinié. Préface de C. Vogt. Paris: P. Reinwald, 1868.

———. *De la variation des animaux et des plantes à l'état domestique.* Traduit sur la seconde édition anglaise par Ed. Barbier, préface de Carl Vogt, Paris: P. Reinwald, 1879.

———. *La descendance de l'homme et la sélection sexuelle.* Traduit de l'anglaise par J.J. Moulinié. Préface par C. Vogt. Paris: P. Reinwald, 1872.

———. *Voyage d'un naturaliste autour du monde fait à bord du navire Le Beagle de 1831 à 1836.* Traduit de l'anglais par Ed. Barbier. (Première édition française) Paris: P. Reinwald, 1875.

———. *The Expression of the Emotions in Man and Animals.* London, 1873.

———. *The Expression of the Emotions in Man and Animals.* New York: Appleton, 1898.

———. *The Formation of Vegetable Mould through the Action of Worms.* New York: MacMillan, 1902.

———. *The Origin of species By Means of Natural Selection: Or the Preservation of Favored Races in the Struggle for Life.* (With additions and corrections from the sixth and last English edition.) Two volumes. New York: Private Edition, 1901.

———. *The Voyage of the Beagle.* New York: Philosophical Library, 1946.

———. *What Darwin Saw in His Voyage Round the World in the Ship* Beagle. New York: Reprint of the 1879 edition.

Darwin, E. *Zoonomia; or, The Laws of Organic Life, in Three Parts.* Two volumes, second American edition prepared from the third London edition. Boston, 1803.

Darwin, E. and . Seward A, editors. *More Letters of Charles Darwin: A Record of His Work in a Series of Hitherto Unpublished Letters.* New York: Private Publication, 1903.

Deacon, T. *The Symbolic Species: The Co-Evolution of Language and the Brain.* New York: Norton Publishers, 1997.

Dechert, C. "Truth, Value, and Intercultural Dialog in an Emerging Global Culture." Contemporary Philosophy. Published by REALIA, Institute for Advanced Philosophic Research, vol. 19:1 and 2, 41. 44-45 (January/February/March/April 1997):

Dennett, D. "Appraising Grace: What Evolutionary Good is God?" *The Sciences.* (January/February 1997): 39–44.

Desmond, A. *Huxley: From Devil's Disciple to Evolution's High Priest.* London: Addison-Wesley Publishers, 1997.

Desmond, A and J. Moore. Quoted by John P. Wiley Jr. q v infra.

Drees, W. *Religion, Science, and Naturalism.* New York: Cambridge University Press, 1996.

Dugatkin, L. "Jealousy's Purpose." Review of David M. Buss's book, *The Dangerous Passion: Why Jealousy Is as Necessary as Love and Sex. Wilson Quarterly.* (Spring 2000): 129.

Editorial. "Kansas Conundrum." *The Washington Times.* (August 19, 1999): A16.

Ehrenreich, B and J. McIntosh. "The New Creationism." *The Nation.* New York: (June 9, 1997). Quoted in *Wilson Quarterly.* (Autumn 1997): 133.

Eissler, K. Interview of Albert Hirst 16 March and 30 March, 1952. In Siegfried Bernfeld Collection, Container 19 (Referred to in texts and catalogues as "Eissler Interview"). Washington, D.C.: Library of Congress.

Ekman, P. Editor of new edition of Charles Darwin's *The Expression of the Emotions in Man and Animals.* Quoted by John P. Wiley Jr. q v infra.

Feldhay, R. *Galileo and the Church.* London: Cambridge University Press, 1997.

Filice, F. (Rev) "The Evolution Wars (Part 1)." Bulletin Catholic

Assoc. of Scientists and Engineers (January 1995): 1–2.

———. "The Concept of Evolution." Bulletin of Catholic Assoc. of Scientists and Engineers (January 1995): 1.

———. Darwin, Part 2. Bulletin of Catholic Assoc. of Scientists and Engineers (February 1996): 1 and 3.

Finch, C. University of Southern California. Quoted by John M. Montgomery q v infra.

Fisher, M. "Most Evil Person of the Millenium." *The Washington Post*. (December 31, 1995).

Freud, S. *The Standard Edition of the Complete Psychological Works of Sigmund Freud,* 24 Volumes, vol. 10, London: Hogarth Press. Original year of publication 1909; subsequent editions 1959 and 1995: 153–320.

Futavama, D. *Evolutionary Biology.* Sunderland, MA: Sinaur Associates, 1942.

Garcia, S. "Sociocultural and Legal Implications of Creating and Sustaining Life through Biomedical Technology." *Journal of Legal Medicine,* vol.17: 4. (December 1996): 469–525.

Gaspari, A. "Darwin Revisited. Inside The Vatican." Document. (January 1997): 469–525.

Gaudiani, V. Editorial, *The Pharos of Alpha Omega Alpha,* vol. 59:4. (1996): 33.

Goldschmidt, T. *Darwin's Dreampond.* Cambridge: M.I.T. Press, 1996.

Goodstein, D. "After the Big Crunch." *Wilson Quarterly*. (Summer 1995): 53–60.

Gould, S. J,. "Let's Leave Darwin Out of It." *The New York Times.* (May 29, 1998): A3.

Green, R. "Human Embryo Research: What Are the Issues?" *Journal of Medical Ethics.* Lahey Clinic, Burlington, MA (Spring 1998).

Hall, M. and S. *The Truth, God or Evolution.* Grand Rapids, MI: Baker Book House, 1975.

Henig, R. *The Monk In The Garden. The Lost and Found Genius of Gregor Mendel, the Father of Genetics.* New York: Houghton Mifflin, 2000.

Hercik, F. and L. Novak. Chybné Zaklady Mendelismu, Cs. Biologie, I. [Erroneous Basis of Mendelism] 1952: 254–62.

Hey, J. and E. Harris. "Modern DNA provides clues to a division in the ancestral tree of human forebears." Washington, D. C.: Proceedings of the Natural Academy of Sciences (February 1999).

Himmelfarb, G. *Darwin and the Darwinian Revolution.* London: Private publication, 1959.

Hirst, A. Autobiography, unpublished. (Referred to as autobiography, not so titled.) Quoted by Lynn, 1997: 23.

Hodgson, S. *The Theory of Practice: An Ethical Enquiry in Two Books.*

London: Longmans, Green, Reader, and Dyer, 1870.

Holmes, D. Research associate of Steve Austed in delaying rate of aging process. Quoted by Montgomery q v infra.

Hubbard, Ron L. *The Rediscovery of the Human Soul.* Personal. Los Angeles: Public Relations Office International (Scientology), 1996.

Hulbert, A and J. Harris. "Raising the American Child." *Wilson Quarterly,* vol. 23:1. (Winter 1999): 19–20.

Huxley, T. *L'évolution et l'origine des espèces.* Avec une préface de l'auteur pour l'édition française. Paris: P. Baillière, 1892.

———. *An Introduction to the Classification of Animals.* London, 1869.

———. *Leçons de physiologié élémentaire.* Traduites de l'anglais sur la troisième éditions par le Dr. E. Dally. Paris: P. Reinwald, 1869.

———. *Lessons In Elementary Physiology,* 6th edition. London, 1879.

Iannaccone, L, R. Stark, and R. Finke. "Rationality and the Religious Mind." *Economic Inquiry.* College Station, TX: Dept. of Economics, Texas A&M University (July 1998).

Inquiry: From a lawyer's point of view. *The Theory of Evolution as Applied to Man.* Chicago: Lakeside Press, 1928.

Irwin, A and B. Wynne B, editors. *Misunderstanding Science?* West London: University of West London Press, 1996.

Jablonski, J. "Ten-year Study of Evolutionary History of 191 Mollusk Lineages." *Nature* (January 1997). (Also quoted by Santos.)

Jacyne, L. Book review of Marshall and Magoun in *Bulletin of History Medicine,* vol 73:2. (Summer 1999): 325–326.

James, W. *The Varieties of Religion Experience: A Study in Human Nature.* Gifford Lectures on Natural Religion, delivered at Edinburgh 1901–1902. London: Longmans, Green and Co., 1902.

Johannesburg AP dispatch reported in *The Washington Times.* (March 12, 1999): A17.

Johnson, H. *The Bible And Early Man.* New York: Declan X. McMullin Co., 1948.

Johnson, P. Quoted by Larry Witham in *The Washington Times.* (January 17, 1996) quid videt infra.

———. "Shaping Up for a New Moral Catastrophe in the Twenty-first Century." *The Spectator,* (October 17, 1998): 26.

———. "The Gorbachev of Darwinism." *First Things.* (January 1998). Quoted in *The Washington Times.* (January 6, 1998): A2.

Jones, E. *The Life and Work of Sigmund Freud,* vol. 2. New York: Basic Books, 1955.

Jones, S. *Almost Like A Whale: The Origin Of Species Updated.* New York: Doubleday, 1999.

Jongejan, F., W. Goff, and E. Comus. "Tropical Veterinary

Medicines: Molecular Epidemiology of Hemoparasites and Their Vectors." Proceedings, New York Academy of Sciences, vol. 849 (1998).

Kafka, P. "God and Man at UCSD." *Forbes.* (February 23, 1998): 39.

Kaminer, W. "The Last Taboo." *The New Republic.* (October 14, 1996).

Karlgaard, R. "Faith and Reason." *Forbes ASAP.* (October 4, 1999): 147–148.

Keane, G. *Creation Rediscovered: Evolution and the Importance of the Origins Debate.* Rockford: Tan Books, 1999.

Keltner, D. and R. Robinson. Psychological Studies conducted at the University of California at Berkeley (Keltner) and Harvard University (Robinson). Quoted by Morin, q v infra.

Kevles, D. "The Crisis of Contemporary Science." *Wilson Quarterly.* (Summer 1995): 41–52.

King-Hele, D. *Doctor of Revolution: The Life and Genius of Erasmus Darwin, 1731–1802.* London: Faber and Faber; 1977.

Kitto, H. *The Greeks.* Toronto, Canada: Pelican Books, Penguin Books Ltd., 1951.

Knight, J. "Creativity and Renewal in the Humanistic Physician." *Arizona Medicine,* vol. 39:3. (1982): 45–47.

Koerner, B. "Out of the African Past." *U.S. News and World Report.* (March 29 1999): 72.

Kragh, H. *Cosmology And Controversy.* Princeton NJ: Princeton University Press, 1996.

Kurlansky, M. *Cod: A Biography of the Fish That Changed the World.* Boston: Walker Press, 1998.

Lakoff, R. "The Growth of the Little Gray Cells." *The Washington Post* (Book World). (November 23, 1997): 6

Lamarck, M de. *Histoire Natrelle des Animaux Sans Vertebres.* Paris: Verdiere Presse, 1815.

Lawson-Tancred, H. *Thoughts from a Dying Star. Almost Like a Whale: The Origin of Species* Updated by Steve Jones. Book review by Hugh Lawson Tancred in *The Spectator.* (July 31, 1999): 26–27.

Levenson, T. *Measure for Measure: A Musical History of Science.* New York: Simon and Schuster, 1995.

Lifton, R. *The Nazi Doctors.* New York: Bais Books, 1986.

Lund, N. Professor of Law at George Mason University. "Biology takes on its own form of morality." Letters to the Editor. *The New York Times.* (June 5, 1998): A22.

Lynn, D. "Sigmund Freud's Psychoanalysis of Albert Hirst." Bulletin of the History of Medicine, vol. 71:1. (Spring 1997): 90–91.

McKee, J. *The Riddled Chain. Chance, Coincidence, and Chaos in Human Evolution.* Rutherford, NJ: Rutgers University Press, 2000.

McNamara, K. *Shapes of Time, The Evolution of Growth and Development.* Baltimore: Johns Hopkins University Press, 1997.

Marcozzi, V. *Darwin versus Christian Faith.* Quoted by Antonio Gaspari. Quid videt supra.

Marscall, L. "Books in Brief: Mr. Science." *The Sciences*. New York: NY Academy of Sciences, (January/February 1998): 44–45.

Marshall, L. and H. Magoun. *Discoveries in the Human Brain: Neuroscience Prehistory, Brain Structure, and Function.* Totowa, NJ: Humana Press, 1998.

Martin, L. "The Big Flap." *The Sciences, v*ol. 38:2. (March/April 1998): 10–12.

Matouskova, B. and O. Matousek. *Darvinism V Tsechoslovakii.* Moskva: Annaly Biologii, 1959: 1, 31–52.

Mayr, E. and W. Provine, editors. *The Evolutionary Synthesis:Perspectives on the Unification of Biology.* Cambridge: Harvard University Press, 1998.

Medawar, P. *Evolutionary Theory of Aging*. Quoted by Doug Stewart q v infra.

Meselson, M. "Evolution without sexual reproduction and genetic recombination." Lecture given at Marine Biological Laboratory, Woods Hole, MA (August 14, 1998).

Michod, R. *Darwinian Dynamics. Evolutionary Transitions in Fitness and Individuality.* Princeton NJ: Princeton University Press, 1999.

Miller, G. *The Mating Mind. How Sexual Choice Shaped theEvolution of Human Nature.* New York: Doubleday, 2000.

Milloy, S. and M. Gough. *Silencing Science.* Washington, D.C.: Cato Institute, 1999.

Mitscherlich, A and F. Mielke; statements by A. Ivy, T. Taylor, and L. Alexander. *Doctors of Infamy: The Story of the Nazi Medical Crimes.* Translation by H. Norden. New York: Henry Schuman, 1949.

Modesto, S. Johannesburg AP dispatch. (March 1999).

Montgomery, J. "High Flight, Hard Shells: Evolution's Clues to Agelessness." *The Washington Post,* Science/Biology Section. (June 22, 1998): A3.

Moorehead, A. *Darwin and the* Beagle. New York: Harper and Row, 1969.

Morin, R. "New Facts and Hot Stats from the Social Scientists." *The Washington Post,* (January 11, 1998): C5.

Mueller, L. Research associate of Michael Rose, quoted by Montgomery q v supra.

National Academy of Sciences. *Guidebook for teachers, parents, school administrators, and policy makers.* Washington, D.C. 1998. Quoted in *The New York Times* quid videt infra.

New York Times. "Scientific Panel on Teaching of Evolution." *The New York Times National* (April 10, 1998): Q14.

Nowers, W. President, American Family Association. "Why Deny the Evidence?" *The Washington Times.* (January 21, 1996): B5.

Ogburn, C. Researcher at University of Washington who with colleagues have evidence that appears to fit with the evolutionary argument on the slowdown of aging. Quoted by Montgomery q v supra.

Orel, A. *Das Weltantlitz. Eine gemeinverstandliche Natur-,Kultur-, Religions-, und Geschichtsphilosophie.* Mainz: Grünewald, 1993: 175.

Orel, V. *Gregor Mendel: The First Geneticist.* Translated by Stephen Finn, New York: Oxford University Press, 1996: 188.

Ospovat, D. *The Development of Darwin's Theory.* New York: Cambridge University Press, 1995.

Ostrom, J. "Bones of Contention." *The Sciences,* vol. 38:5. (September/October 1998): 3

Padian, K. "Peer Review: Letters From Readers." *The Sciences,* vol. 38:5. (September/October 1998): 3, 9.

Panush, R. Upon finding a Nazi anatomy atlas: "The lessons of Nazi Medicine." *The Pharos of Alpha Omega Alpha,* vol. 59:4. (1996): 18–23.

Paul, L. Associate professor of psychology at Temple University. "Letters to the Editor." *The New York Times.* (June 5, 1998): A22.

Plotkin, H. *Evolution in Mind: An Introduction to Evolutionary Psychology.* Cambridge: Harvard University Press, 1998.

Polkinghorne, J. "An Intelligible Universe." *Commonweal.* August 16, 1996.

———. *Beyond Science.* London: Cambridge University Press, 1996.

Pope John Paul II. "Message to Pontifical Academy of Sciences." *L'Osservatore Romano,* no. 30, 44. (October 1996): 3, 7.

———. Science serves humanity when joined to conscience. Words spoken to attendees at the International Conference on Space Research, University of Padua on January 11, 1977. *L'Osservatore Romano,* no. 4 (1475). (January 22, 1997): 1.

Popeo, D. Chairman, Washington Legal Foundation. *Forbes.* (February 22 1999): 36.

Proctor, R. *Racial Hygiene. Medicine Under the Nazis.* Cambridge: Harvard University Press, 1988.

Purcell, R. *Special Cases: Natural Anomalies and Historical Monsters.* New York: Chronicle Books, 1997.

Raby, P. *Bright Paradise: Victorian Scientific Travellers.* Princeton: Princeton University Press, 1997.

Ramachandran, V. Director of Brain and Cognition Center, University of California at San Diego. Quoted by Kafka q v supra.

Raugust, B. "Scientific Panel Pushes Teaching of Evolution in Public Schools." *The New York Times.* (April 10, 1998): A14.

Reeves, T. "Not So Christian America." *First Things.* New York: Institute on Religion and Public Life (October 1996).

Reich, D. "Science and Technology." *Wilson Quarterly.* (Winter 1997): 103–104.

Richter, O. *75 Jahre Seit Mendels Grosstat und Mendels entdeckungen. Verhandlungen des Naturforschenden Vereines.* Brünn: 1941: 72, 109–73.

Rittmann, J. "Curiouser and Curiouser." *The Sciences.* New York: NY Academy of Sciences (May/June 1998): 39–42.

Ritvo, H. *The Platypus and the Mermaid and Other Figments of the Classifying Imagination.* Boston: Harvard University Press, 1997.

Roberts, J. *History of the World:* 700–705. Quoted by Sandra Anderson Garcia q videt supra.

Robinson, V. *The Story of Medicine.* New York: The New Home Library, 1931: 38, 58, 389–91.

Roche, G. *World without Heroes: The Modern Tragedy.* Hillsdale, MI: Hillsdale College Press, 1987: 233–249.

Rodenhauser, P. "On Creativity and Medicine." *The Pharos of Alpha Omega Alpha,* vol. 59:4. (May/June 1998): 2–6.

Rose, M. Population geneticist at University of California at Irvine. Quoted by Montgomery q v supra.

———. *Darwin's Spectre. Evolutionary Biology in the Modern World.* Princeton: Princeton University Press, 1998.

Ruben, J. "Letters from Readers." *The Sciences,* vol. 38:5, (September/October 1998): 3, 49.

Rubidge, B. Quoted in Johannesburg AP dispatch, March 1999.

Ruse, M. *Monad to Man: The Concept of Progress in Evolutionary Biology.* Cambridge: Harvard University Press, 1997.

Santos, L. "Downsizing Evolution." *The Sciences.* (March/April 1997): 46.

Segerstrale, U. *Defenders of the Truth. The Battle for Science in the Sociobiology Debate and Beyond.* London: Oxford University Press, 2000.

Seward, A. *Darwin and Modern Science. Essays in Commemoration of the Centenary of the Birth of Charles Darwin and of the Fiftieth Anniversary of the Publication of* The Origin of Species. Cambridge: Cambridge University Press, 1909.

Shafer, G. Darwin centenary publications. *Eleventh Report of the Michigan Academy of Science.* East Lansing, Michigan Academy of Science, 1909.

Shute, E. *Flaws in the Theory of Evolution.* Nutlye, NJ: Craig Press, 1962.

Siemann, E., G. Tilman, and J. Haarstad. Department of Ecology, University of Minnesota, St. Paul. "Cope's Rule and Insects." *Nature.* (April 1996).

Simpson, G. *The Major Features of Evolution.* New York: Columbia University Press, 1969.

SJG (Stephen Jay Gould). Letters to the Editor. *The New York Times.* (May 29, 1998.) Quoted by Nelson Lund, q v supra.

Sober, E and D. Wilson D. *Unto Others: The Evolution and Psychology of Unselfish Behavior.* Cambridge: Harvard University Press, 1998.

Soller, M. "Kissing Cousins?" *The Sciences,* vol 38:2. (March/April 1998): 5–8.

Stein, R. "Eyes May Have It: A Common Origin." *The Washington Post,* (March 18, 1997): A3.

Stewart, D. *Solving the Aging Puzzle.* Washington, D.C.: Smithsonian Institution, vol. 28:10. (January 1998): 37–44.

Storey, K. and J. Storey. "Lifestyles of the Cold and Frozen." *The Sciences.* (May/June 1999): 33–37.

Taylor, E. *Biological Consciousness and the Experience of the Transcendent: William James and American Functional Psychology.* Bethesda, MD: National Library of Medicine and Washington, D.C.: American Psychological Association, Joint Publication, 1992: 53.

Thomson, J. *Darwinism and Human Life.* The South African Lectures for 1909. New York: Holt, 1910.

Thornhill, R. and C. Palmer. "Why Men Rape." *The Sciences.* (January/February 2000). New York: NY Academy of Sciences. Based upon original publication, *A Natural History of Rape.*

Troisi, A and M. McGuire. *Darwinian Psychiatry.* New York: Oxford University Press, 1998.

Viviani, F and G. Casagrande, editors. *Physical Activity and Health: Physiological, Behavioral, and Epidemiological Aspects.* Padua: Unipress, 1998.

Waddington, C. *Theories of Evolution: In a Century of Darwin.* S. A. Barnette, editor. London: Henemenn Press, 1958.

Weiner, J. *Time, Love, Memory.* New York: A. A. Knopf, 1999.

Wiley, J. Jr. "Expressions: The Visible Link." *Smithsonian Magazine,* vol. 29:3, not paginated. (June 1998).

———. "Phenomena, Comments and Notes." *Smithsonian Magazine,*vol. 26:9. (December 1995): 26–28.

Willey, B. *Darwin and Butler, Two Versions of Evolution.* The Hibbert Lectures, 1959. New York: Harcourt-Brace, 1960.

Wilson Quarterly. "The Left's Creationists." Abstract on *The New Creationism* by Barbara Ehrenreich and Janet McIntosh in *The Nation.* (June 9, 1997): 133. *Wilson Quarterly* (Autumn 1997).

Wilson, E. *Consilience: The Unity of Knowledge.* New York: Alfred A. Knopf, 1998.

———. "Resuming the Enlightenment Quest." *Wilson Quarterly.* (Winter 1998): 16–29.

Wilson, F. *The Hand: How Its Use Shapes the Brain, Language, and Human Culture.* New York: Pantheon Books, 1998.

Winchester, S. *The Professor and the Madman. A Tale of Murder, Insanity and the Making of the Oxford English Dictionary.* New York: Harper Collins, 1998.

Witham, L. "Creation-Evolution Debate Taking on a Less-Shrill Debate." *The Washington Times.* (January 17, 1996): A2.

Woodrow Wilson Quarterly (author anonymous). "The Biological Great Gatsby." Arts and Letters, vol. 23:2 (Spring 1999): 114.

Wozniak, R. *Mind and Body: René Descartes to William James.* Bethesda, MD: National Library of Medicine and Washington, D.C.: American Psychological Association, Joint Publication, 1992: 10.

Glossary

* Dr. Samuel Johnson was a historic English author, conversationalist, and scholar 1709–1784. He had regional tuberculosis.

*(21) Neogenesis according to Rev. Teilhard De Chardin, S.J. *(Social and Cultural Implications)* means the development of mind or psychic state. As a paleontologist, Teilhard traced the evolution of human consciousness, freedom, and information, key requirements for the growth of society, culture, and human community.

** Sybaritic is objective for pleasurable as derived from a sybarite, who luxuriates in the good life according to his/her own designs.

* (26) Freud's conclusion was founded upon Lamarck's principle of the inheritance of acquired characteristics that included behavior patterns without any mention of morality and ethical conduct (Jones 1955).

* (28) Sir Charles Bell was a Scottish physiologist in London (1774–1842) whose name is eponymous. In medical literature there is Bell's law, Bell's nerve, Bell's palsy, and Bell's phenomenon. Bell's law is also known as Bell-Magendie law (François Magendie, French physiologist, 1783–1855, was the pioneer of experimental physiology in France), which is "the anterior roots of the spinal nerves are motor roots and the posterior are sensory roots." Bell's nerve is the long thoracic nerve. Bell's palsy is a peripheral facial paralysis due to a lesion of the facial nerve (seventh cranial nerve) which results in characteristic distortion of the face. Bell's phenomenon is an outward and upward rolling of the eyeball on the

attempt to close the eye: it occurs on the affected side in peripheral facial paralysis.

** (28) Ivan Petrovich Pavloff (1849–1936), also known as Pavlov, was a Russian physiologist whose primary contribution concerned conditioned reflexes producing salivary and gastric secretions.

* (66) Ernest Heinrich Haeckel (1834–1919) was a German biologist and philosopher whose early espousal of Darwinism led him to found upon the evolutionary hypothesis a materialistic monism which he advanced in his writings. His most popular work was *The Riddle of the Universe.* Succinctly, monism is metaphysically the view that there is but one fundamental reality. Christian Science is an example of a popular contemporary religion built on an extreme monistic theory of reality. Epistemologically, monism holds that the real object and the idea of it (perception or conception) are one in the knowledge relationship of each.

* Hydrogen, a gaseous element is the most abundant element in the universe (atomic number 1). Highly flammable, it is used in rocket fuels. Hydrogen supposedly emanated from a volcano during a violent, fire-filled eruption which was part of the big-bang that brought about the universe.

* Paralogism is false reasoning (falsity).

** (69) Solecism is any impropriety, a mistake, an incongruity.

* (74) Bruno refers to Saint Bruno of Cologne (1030?–1101) an outstanding German educator; founder of the Carthusian order of monks.

**Argumentum ad hominen* is one directed to the person rather than to the subject. An argument appealing to personal prejudice and emotion rather than to logic or reason.

* The missing link is a theoretical primate postulated to bridge the evolutionary gap between the anthropoid apes and mankind. The word *missing* is used to express that something is lacking which is needed to complete a series. To substantiate Darwinism this missing link is essential to elevate evolution from a theory to a scientific fact.

* Burden of proof is a legal term pertaing to substantial facts needed to substain alligations in a court of law and/or an adjudication hearing.

* (106) The Human Genome Project (HGP) is a $3.5 billion federally funded enterprise involving thousands of scientists. It is a fifteen-year guaranteed research endeavor.

* (108) Henry Louis Mencken (1880–1956) was an American editor, critic, and author who wrote on proper American language.

* (109) Memes are complex ideas—in the sense of something one might patent or copyright, not in the sense of an element of experience. The idea of the wheel is a meme. So is the idea of hijacking airplanes—a meme that tends to replicate even though it is not beneficial to people. This definition is by Daniel C. Dennett, who is distinguished in arts and science. He is the professor and the director of the Center for Cognitive Studies at Tufts University in Massachusetts. Walter Burkett of Harvard University authored the book *Creation of the Sacred: Tracks of Biology in Early Religions.* According to Dennett, "Burkett abruptly dismisses memes as mere metaphor and hence tends to overlook the possibility of radically different answers to the question of beneficiary. As a consequence, he sometimes fails to notice when his attempts at explanation wander between quite distinct possibilities, mixing considerations of varying relevance, and ignoring the disparate implications of his hypotheses." This same criticism is applicable to Darwin.

*The National Academy of Sciences in 1976 asserted that science and religion are mutually exclusive. Many non-Christian religions believe in evolution and therefore the motivation behind the Academy's position is the desire to deny Christianity (Nowers).

*(6,112) Thomas Henry Huxley (1825–1895) was a British biologist and advocate of Darwinism; grandfather of Aldous Leonard Huxley (1894–1963) an English novelist and critic who resided in America as well as grandfather to Sir Julian Sorell Huxley (1887–1975) a British biologist and author. Thomas Henry Huxley's father was a poor schoolmaster who could not afford to pay Thomas's medical school tuition.

*(117) Pius XI (Achille Ambrogio Damiano Ratti; 1857–1939), reigned from 1922 to 1939. He signed the Lateran Treaty in 1929 with Benito Mussolini, Prime Minister of Italy, which gave autonomy to the modern Vatican city-state.

*Clarification of the Pope's statement. The English language edition of the Vatican newspaper, *L'Osservatore Romano,* has pointed out a discrepancy in its translation of a message by Pope John Paul II on evolution. In his message to the Pontifical Academy of Sciences on October 23, 1996, the pope said that over the last fifty years new knowledge has emerged that shows the theory of evolution to be "more than a hypothesis." His point was that evolution was now accepted by a wide range of scientific disciplines doing independent research. In the English language *L'Osservatore,* however, the Pope's sentence was translated as meaning that new knowledge has "led to the recognition of more than one hypothesis in the theory of evolution."

* Pius IX's whose original name was Giovanni Maria Mastai-Feretti (1792–1878), reigned from 1846–1878. He proclaimed the dogmas of the Immaculate Conception and papal infallibility.

* (127)*Agora* is a meeting place, an assemblage, a marketplace for discussion.

** (127) Relativism is the theoretical view that truth is relative and may vary from individual to individual, from group to group, or from time to time, having no objective standard. Relativism in regard to epistemology is the theory that all human knowledge is relative, a polyadic propositional function, to the knowing mind and to the conditions of the body and sense organs. Relativism adheres to its dictum that truth is subjective entirely. In epistemology subjectivism is the restriction of knowledge to the knowing subject and its sensory, affective, and volitional states. External things are inferred from the subjective viewpoint and interpretation thereof. In the study of the nature of values and value judgments (axiology) the subjectivism doctrine is that moral and aesthetic values represent the subjective feelings and reactions of individual minds and have no status independent of such reactions.

*An explanation of "classic configuration" is needed. This two-word alliteration is a designation for biography, biobibliography, and history of ideas. Designed by John C. Burnham, it is presented in his essay, "How the Idea of Profession Changed the Writing of Medical History." Published in the *Medical History Supplement*, No 18. London: Wellcome Institute for the History of Medicine, 1998.

* (135) Chevalier La Marck was the name of Jean Baptise Pierre Antoine de Monet (1744–1829), a well-known French naturalist. He invented the word *biology* defining it as the study of living things: animate things in contrast to what is inanimate.

* Niles Eldredge is a curator at the American Museum of Natural History in New York who is quoted by Witham. As an evolutionary theorist, he has no conflict in accepting human descent from apes.

* (139) Phillip Johnson, a legal scholar, is a leading public critic of evolution based on his 1991 book, *Darwin On Trial*. He is a professor at the University of California at Berkeley.